The Manual of Below-Grade Waterproofing

The ever evolving technology of waterproofing presents challenges and risks for architects and engineers who do not specialize in the field. The revised edition of *The Manual of Below-Grade Waterproofing* provides the education and product information to enable designers to take a sound, fundamental approach to these contemporary challenges.

Building designers specify waterproofing systems and materials that are often based on limited and subjective manufacturers' literature or past experience with systems that work under specific conditions, but will fail in other installations. Leakage usually leads to litigation. This book gives you the tools to prevent that.

This manual covers the history and science of waterproofing materials, the considerable distinctions between waterproofing roofs and plazas and below-grade surfaces, the critical procedures for protecting waterproofing materials during construction, diagnosing and remediating leaks, writing specifications, and detailing waterproofing components. The pros and cons of every waterproofing material and system are comprehensively covered. You will learn how to:

- weigh up positive- versus negative-side waterproofing systems;
- weigh up dampproofing versus waterproofing;
- coordinate with all the professionals in the waterproofing delivery chain; and
- follow environmental protection and government regulations.

This book is an essential resource for architects, civil engineers, contractors, designers, materials manufacturers, and all other professionals involved with the design and construction of underground spaces.

Justin Henshell, FAIA, CSI, FASTM, is a partner in Henshell & Buccellato, Consulting Architects, in Shrewsbury, New Jersey, USA. He has been practicing architecture since 1952 and has specialized exclusively in moisture-related problems in the building envelope internationally since 1974. He has lectured, and written more than 60 papers about waterproofing.

The Manual of Below-Grade Waterproofing

Second edition

Justin Henshell

LONDON AND NEW YORK

First published in paperback 2024

First edition published 1999 by Wiley

Second edition published 2016
by Routledge
4 Park Square, Milton Park, Abingdon, Oxon OX14 4RN

and by Routledge
605 Third Avenue, New York, NY 10158

Routledge is an imprint of the Taylor & Francis Group, an informa business

Publisher's Note
The publisher has gone to great lengths to ensure the quality of this reprint but points out that some imperfections in the original copies may be apparent.

British Library Cataloguing-in-Publication Data
A catalogue record for this book is available from the British Library

Library of Congress Cataloging in Publication Data
Names: Henshell, Justin, author.
Title: The manual of below-grade waterproofing / Justin Henshell.
Description: 2nd Edition. | New York : Routledge, 2016. | Includes bibliographical references and index.
Identifiers: LCCN 2016001397 | ISBN 9781138668195 (hardback : alk. paper) | ISBN 9781315618753 (ebook : alk. paper)
Subjects: LCSH: Dampness in buildings--Prevention. | Waterproofing. | Foundations--Design and construction.
Classification: LCC TH9031 .H46 2016 | DDC 693.8/92--dc23LC record available at http://lccn.loc.gov/2016001397

ISBN: 978-1-138-66819-5 (hbk)
ISBN: 978-1-03-292279-9 (pbk)
ISBN: 978-1-315-61875-3 (ebk)

DOI: 10.4324/9781315618753

Typeset in Sabon
by FiSH Books Ltd, Enfield

Contents

Acknowledgements

I wish to thank all those who helped me bring this book to fruition. Special thanks go to my partner Paul Buccellato, AIA who meticulously prepared and organized the illustrations and Richard Fricklas who reviewed the entire manuscript and provided thought-provoking and incisive comments.

I owe a special debt to Dorothy Lawrence[1] of Laurenco Waterproofing Systems and Charles O. Pratt,[2] PE, Consultant who fanned my interest in waterproofing, held my hand, offered encouragement and patiently and cheerfully tutored me with their wealth of technical and historical information.

Thanks also to my secretary Margaret Mary McCarthy, who spent countless hours in word processing and proofreading the manuscript. Among the many who helped with technical input are Carl G. Cash,[3] PE; Simpson, Gumpertz & Heger; Wayne Tobiasson, PE, Gerald Zakim, GZA; Tim Eorgin, Carlisle Coatings & Waterproofing; Donald Comerford, Waterproofing Systems of New Jersey and Alfred Kessi, Vandex Sales & Services, Inc. (now Aquafin)

Last, but not least, my sincere appreciation and gratitude to C. W. "Bill" Griffin, who edited the first edition and who organized the manuscript, put technical jargon into understandable English, challenged me at every turn and rewrote large portions of the text.

For the second edition, I am indebted to Thomas Planert of SsesCo who guided me through the chemistry of chemical injections, Stacy Byrd of CETCO and Ray Wetherholt who edited the bentonite chapter and waterstops, Richard Boone the guru of roofing chemistry who contributed his vast knowledge to Chapter 11, and all who provided suggestions for improving the manuscript for the second edition.

Finally, thanks to John Henshell who edited the second edition and exceeded the standard for research and writing clarity.

Notes

1 Deceased.
2 Ibid.
3 Ibid.

About the author

Justin Henshell, FAIA, CSI, FASTM, is a partner in Henshell & Buccellato, Consulting Architects, based in Shrewsbury, New Jersey. He has been practicing architecture since 1952 and has specialized exclusively in moisture-related problems in the building envelope nationwide since 1974.

Mr. Henshell is past chairman of the ASTM sub-committee on Roofing Membrane Systems and serves on the sub-committee on Waterproofing and Dampproofing Systems where he was the principle author for the ASTM Standards D 5898 Guide for Standard Details for Adhered Sheet Waterproofing and D7832 Performance Attributes of Waterproofing Membranes Applied to Below-Grade Walls/Vertical Surfaces (Enclosing Interior Spaces).

He has lectured on roofing and waterproofing in North America and Europe and authored over 60 papers on those subjects.

Among more than 100 below-grade waterproofing projects on which the firm has consulted are: Whitney Museum of American Art, New York; Wharton School of Business, University of Pennsylvania; Transbay Transit Center, San Francisco; Terminal C, Dubai Airport; Hudson Yards, New York; Glenstone Museum, Potomac, MD; Memorial Sloan Kettering, New York; and Museum for Westward Expansion, St. Louis.

About the editor

John Henshell (john@johnhenshell.com) is Justin Henshell's son. He is a strategic writer/editor/communications consultant with 30 years of experience. He adds value to his clients' words through adept use of diction, syntax, context, and visual images.

John Henshell has written Web content, articles, newsletters, marketing collateral, buyers' guides, advertising copy, brochures, features, press releases, business plans, reports, and technical articles. His clients have included *PC World*, ConsumerSearch.com, the Oregon AeA magazine, EBSCO, The Center for Digital Ethics and Policy at Loyola University Chicago, *Computer Bits*, goodguys.com, the *Journal of Information Ethics*, and business.com.

1 Introduction

A. What is waterproofing?

In this manual, in accordance with ASTM usage, *waterproofing* means a system designed to resist hydrostatic pressure exerted by moisture in a liquid state. It is differentiated from dampproofing, which is designed merely to resist the flow of moisture in a gaseous state (i.e., water vapor). See Chapter 3 "Dampproofing," and Chapter 5 "Waterproofing," for a detailed explanation of this distinction.

Waterproofing is often erroneously defined as a coating applied to above-grade masonry or concrete walls; a membrane cover on spandrel beams; a membrane in split-slab construction under a mechanical or shower room; a coating on a parking garage, in a tank, or pool; as well as four different treatments of below-grade foundation walls and slabs.

A subtler source of semantic confusion is distinguishing between a waterproofing system and a low-slope roof system when the roof system is a so-called Protected Membrane Roof (PMR) assembly. In a PMR, the membrane is placed directly on the deck below the insulation, instead of in its conventional, weather-exposed location atop the insulation. When a PMR is ballasted with concrete pavers, it resembles a waterproofed plaza. A roof plaza can, in fact, be identical to a waterproofed plaza at or below grade.

The distinguishing feature differentiating a PMR from a waterproofed plaza system is the accessibility of the membranes in the event of failure. This feature is the basic determinant of manufacturers' willingness to issue guarantees. Manufacturers will generally guarantee a PMR if pavers are installed on pedestals, in which case the membrane is accessible. Most manufacturers will guarantee PMR systems in which pavers, serving as ballast or maintenance walkways, are loose-laid on the insulation. (In industry semantics, such a system is not considered to be plaza waterproofing.) If, on the other hand, the membrane is inaccessible – with wearing surface-units (such as brick, tile, or stone) installed in a mortar setting bed or in sand over a concrete protection slab – then the system is classified as waterproofing, and manufacturers will not guarantee it.

A guarantee should, however, be a secondary consideration. You should select a PMR system only if your primary goal is to provide a client with the dubious assurance of a manufacturer's warranty. If a lower-risk system assuring long service life in a continuously moist environment over a moisture-sensitive space is your primary goal, then a waterproofing system is a more prudent selection.

Another source of semantic confusion is distinguishing between a *below-grade* waterproofing system and a *plaza waterproofing* system. *Though both are used below grade, very few plaza waterproofing systems are designed to resist significant hydrostatic pressure.*

B. Using this manual

You do not need to read this manual cover-to-cover to gain practical guidance. The table of contents and index (or keyword search, if you are reading an eBook edition) should direct you to the material that will help with your current need. Each chapter covers individual waterproofing topics in depth. Cross-references are provided within chapters to direct you to additional relevant material.

The objective of the manual is to help you avoid waterproofing system failures. Manufacturers of roof waterproofing systems provide considerably more selection, design, and installation support than manufacturers of below-grade waterproofing systems do. This manual is dedicated filling that gap as well as providing a sound, fundamental approach to waterproofing problems.

C. Manual's scope and limitations

This manual focuses on waterproofing of below-grade, habitable building spaces subject to hydrostatic pressure, as well as dampproofing. It is therefore limited to covering waterproofing and dampproofing for the following underground building components:

- structural concrete slabs with a wearing surface or earth fill;
- concrete slabs-on-ground below-grade elevation; and
- foundation walls.

Waterproofing (and dampproofing) for the following structures are not covered:

- water-containment vessels;
- vehicular, pipe and similar tunnels not enclosing habitable space;
- above-grade mechanical room floors (not exposed to groundwater or soil chemicals);
- traffic toppings or traffic-bearing waterproofing systems for vehicles

and balconies offering short-term protection, but not resistance to hydrostatic pressure; and
- green vegetative roofs.

This manual is not a substitute for the professional expertise required to design a waterproofing system. It illustrates general design principles. As most waterproofing projects have unique aspects, you will not simply be able to copy details provided here. You will need to adapt them to your project. In some cases, details provided by manufacturers will be more relevant than the general details included in Chapter 17.

Because waterproofing technology is rapidly changing, some systems presented herein may be obsolete or superseded by others as you read this.

D. The importance of waterproofing

Urban planning critics, who discuss the purely architectural aspects of plaza design with informed expertise, normally ignore the functional aspect: The unseen waterproofing system required to sustain these aesthetic triumphs. Space under the street-level plaza, located on prime land sometimes priced at four digits per square foot, is inevitably dedicated to high-level occupancies in which the slightest leakage is intolerable.

A waterproofing system should ideally last the full service life of the building. Thus, durability is the first principle of waterproofing design. Most building systems can be designed for service lives far short of projected building service life. Air-conditioning equipment, lighting, office partitions, communications, even elevators, curtain walls, and roofs can be designed for anticipated replacement as they either become obsolete or wear out under constant weathering. A waterproofing system, however, is like a foundation and structural framework in its need for endurance.

The importance of durability is illustrated by industry practices. Few waterproofing manufacturers offer guarantees or useful warranties, yet manufacturers usually do support roofing materials. The few warranties available are for severely limited liability, for two major reasons:

1. Waterproofing for foundations and under slabs-on-ground is often inaccessible.
2. Replacement of waterproofing systems under plazas may require removal and replacement of tons of overburden or jackhammering removal of concrete protection slabs and concrete topping.

Such drastic remedial action sometimes costs more than 300 times the membrane's cost. That risk motivates prudent designers to ignore first-cost economy when it entails *any* risk of failure. Prudent designers select systems and materials that will work, regardless of a few dimes per square foot cost increase. They also require:

- approved applicators for installing waterproofing systems;
- rigorous quality assurance (QA) inspection programs during installation; and
- flood-testing, whenever practicable.

While the problems of modern waterproofing have multiplied, advances in waterproofing technology have multiplied at an even faster rate, giving you a vastly expanded selection of new systems and materials. Modern polymer chemistry expands your arsenal of weapons against leakage while complicating the problem of selecting the proper combination of systems or materials.

E. The boom in demand for waterproofing

The growing demand for a greater quantity and higher quality of waterproofing in building construction has a simple explanation. The rapid proliferation of air-conditioning following World War II made windowless underground spaces feasible for human occupancy. Before that, underground spaces were severely limited by the practical inability to control interior temperature, humidity, and circulating air quality for comfortable, healthy habitation. The advent of air-conditioning created a demand promoted by many interrelated factors:

- space needs for expanded mechanical plants
- space needs for windowless functions
- trend toward water-shedding curtain walls
- sites with poor drainage
- incentive zoning for open, street-level plazas
- rise in HVAC energy costs
- rising cost of waterproofing failures

Space needs for expanded mechanical plants grew with the demand for sophisticated HVAC systems, electronic data processing centers, and other new technologies in the postwar building boom. The rudimentary building mechanical equipment preceding the 1950s seldom required deep basements. Huge chillers and other mechanical equipment for office skyscrapers created a demand for more underground space. Basements, sub-basements, and even sub-sub-basements were excavated to accommodate this equipment (see Figure 1.1).

Space needs for windowless functions include many functions best conducted in an environment usually isolated from the outside world. As underground locations are necessarily windowless, they are prime locations for a broad spectrum of occupancies: theaters and similar audio-visual uses, archives, mainframe computers, electronic switch gear, parking, retailing, and bowling alleys.

Figure 1.1 This deep excavation, for the New Academic Building at Cooper Union in New York, provided spaces for an auditorium and laboratories below grade where daylight was unnecessary. The groundwater level was 12 feet above the basement floor

The trend toward water-shedding curtain walls aggravated building waterproofing problems. The traditional masonry walls of pre-World War II construction absorbed substantial quantities of rainwater and shed it by evaporation in a cyclical wetting-drying process. Glass and metal wall panels shed this water rapidly via vertical flow directly to the foundation soil. Accumulation of storm water at the foundation thus magnifies drainage and waterproofing problems at the building basement.

Sites with poor drainage became more common during the 1950s when the accelerated tempo of postwar building made prime land scarcer. Foundation soil at previously occupied sites was sometimes contaminated, particularly in reclaimed swamps. Contaminants included acid and alkaline water, insecticides, soil poisoners, and fertilizers and petroleum products discharged from vessels, refineries, and underground tanks.

Zoning regulations promulgated to induce developers to provide more open space in city centers, magnified waterproofing problems. Incentive zoning offered developers greater building heights and floor area in return for street-level plazas. To maximize the economic potential of high-priced urban land, underground commercial space and parking garages were built below these plazas. Waterproofing of these city-center plazas requires much greater dependability and durability than traditional membrane waterproofing of sidewalk vaults over cellar utility spaces.

The energy crises of the 1970s indirectly complicated waterproofing design. Prior to the sudden increases in energy costs, when heating oil cost less than 10 cents per gallon, thermal insulation was invariably omitted from waterproofing systems. Earth-covered buildings, which enjoyed a brief vogue in the early 1970s, are more energy-efficient than conventional, above-ground buildings. Increasing the underground space in a conventional building increases the overall energy efficiency of the building, but presents a tougher waterproofing problem.

Rising costs of waterproofing failures have accompanied modern building trends. Failure may occur even without leakage of liquid moisture into the occupied space. Environmentally sensitive occupancies are routinely located underground in contemporary buildings. Electronic equipment in underground computer rooms and public assembly spaces containing audio-visual equipment require rigorous humidity control as well as leakproof waterproofing. Auditoriums with wood floors are nearly as demanding as computer rooms. Such uses create unprecedented waterproofing problems.

F. Historical background

Paradoxically, the archetypal waterproofing project, the Hanging Gardens of Babylon, anticipated modern waterproofing problems nearly 26 centuries ago, six centuries before the Christian era. One of the seven wonders of the ancient world, the Hanging Gardens featured terraces rising 75 feet from a colonnade superstructure. Waterproofing consisted of bitumen and lead. Along with other plants, trees were planted in the overburden over the earth-covered deck and watered by slave-powered irrigation machines lifting water from the nearby Euphrates (see Figure 1.2).

By the beginning of the twentieth century, waterproofing projects involved more mundane uses, primarily tunnels, dams, pools, and other water-containment structures. Cellar vaults under sidewalks were also protected, normally with built-up coal tar pitch membranes. These early built-up bituminous waterproofing membranes comprised alternate layers of cotton or burlap, organic felts, and coal tar pitch. Membranes were then covered with bricks dipped in hot pitch. Before World War I, built-up waterproofing was applied to building foundations, vertically to brick and tile walls, horizontally to mud slabs in hot coal tar pitch or asphalt. Standard specifications for these early built-up membranes comprised four to six plies for waterproofing and three to four plies for dampproofing. Concrete foundations and floor slabs were cast over these hot-applied membranes. For shallow cellars, a sub-slab drainage system was specified in conjunction with membranes.

Figure 1.2 The Hanging Gardens of Babylon, built by King Nebuchadnezzar six centuries before the Christian era, is the archetypal waterproofing project. Earth-covered terraces planted with trees and other plants were supported on a series of 75-foot-high arches. Irrigation from the Euphrates watered the trees

Because waterproofing membranes were expensive to install and difficult to repair, cementitious waterproofing was introduced a century ago as a less expensive, more convenient alternative. Cementitious waterproofing coats interior (dry) faces of walls and slabs. The most popular of several proprietary compounds was a hydrolithic metallic system containing iron fillings. Oxidation of the filings surfaces, in the presence of mixing water hydrating the cement, expanded the iron fillings' volume to several multiples of its unoxidized volume, and this expanded volume filled interstices in the concrete matrix and formed a tight, water-resistant solid.

Two systems and two materials dominated traditional waterproofing: wet-face (*positive*-side) waterproofing featuring built-up membranes and dry-side (*negative*-side) waterproofing featuring cementitious coatings. Use of these traditional materials has declined to minor status, buried under an avalanche of newer and better materials.

Built-up coal tar pitch membranes survive because almost all contemporary waterproofing materials are incompatible with coal tar pitch. This makes repair of pitch membranes a nightmare unless the same bitumen is used. VOC restrictions and air pollution regulations that ban hot kettles add further complications.

G. Modern waterproofing materials

The development of new materials to replace traditional built-up membranes and cementitious coatings accompanied an accelerated evolution of waterproofing that has also largely replaced drainage media and waterstops with new, superior plastics. A host of modern membrane materials have supplanted hot-mopped bituminous membrane materials in use since the mid-nineteenth century: rubberized-asphalt membranes, single-ply sheets of butyl and Polyvinyl chloride (PVC), one- and two-component liquid-applied membranes (LAMs) and bentonite sheets. The old cementitious coatings with iron or aluminum filings are being replaced by crystalline/chemical conversion coatings capable of sealing tiny shrinkage cracks. Bentonite clay is the basis of many prefabricated composite components. Bentonite-based laminates may contain one or more combinations of high-density polyethylene, geotextiles, and butyl sheets, plus several different versions of kraft paper containers.

By the 1990s, environmental constraints became a significant factor shaping the evolution of waterproofing systems and materials. With the advent of EPA and VOC (volatile organic content) regulations setting TLV (threshold level values promulgated by OSHA), systems releasing volatile solvents and potentially carcinogenic fumes were severely restricted or banned in densely populated center cities. This trend has severely reduced the use of hot-mopped built-up bituminous membranes and promoted their replacement with environmentally harmless systems: liquid-applied membranes, cold-applied systems, and water-based (instead of solvent-based) primers.

Prefabricated modified bitumens (i.e., rubberized asphalt) membranes, supported on HDPE (high-density polyethylene) sheets, are prominent among the replacements for traditional hot-mopped built-up membranes. In addition to reducing (or even eliminating) the polluting fumes emanating from traditional hot-mopped application, modified bitumen membranes represent a leap forward in the goal of building construction: to shift as much work as practicable from field to factory. This process facilitates quality control, labor productivity, and safety. One waterproofing manufacturer, W.R. Grace, introduced a proprietary, self-adhering modified-bitumen membrane, which Grace called "rubberized asphalt," back in the 1960s. Grace's Bituthene had monopolized this market until its patents expired in the early 1990s when several manufacturers began marketing similar products (see Figure 1.3). All modified-bitumen waterproofing membranes marketed today contain an SBS (styrene butadiene styrene) plasticizer. Because it is unreinforced (but laminated to HDPE), SBS-modified bitumen provides superior waterproofing. The alternative roofing products, APP (atactic polypropylene) and SBS-modified bitumen sheets, are reinforced with glass or polyester felts. They have not achieved a significant market share in the waterproofing field because, at exposed edges, the reinforcement can wick water into the sheet, delaminating it and shortening its service life.

Figure 1.3 Self-adhering modified bitumen membranes, with high-density polyethylene (HDPE) facer sheets, are among the most popular replacements for traditional hot-mopped, built-up membranes. The latter are increasingly being banned by anti-pollution ordinances. Pictured is the roof of a below grade gymnasium adjoining a church in lower Manhattan, New York. The membrane will be covered with two feet of earth and landscaped

Single-ply sheets, both thermosetting (i.e., vulcanized rubber) and thermoplastic, have been vastly improved. Butyl is favored over EPDM and neoprene, which is no longer manufactured in large sheets. Butyl has much lower water absorptivity than either EPDM or neoprene, a tremendous advantage for continuously wet waterproofing membranes.

Among the thermoplastic waterproofing sheets, glass fiber-reinforced PVC, in gauges up to 120 mils, has emerged as the material of choice, replacing CPE and unreinforced PVC, whose thinner cross sections often resulted in plasticizer loss and consequent embrittlement as the material aged.

Long used under slabs and for foundations, especially as blindside waterproofing, bentonite clay has become one of the most widely used positive-side waterproofing materials. It comes in a wide variety of laminates, including geotextiles filled with bentonite granules and high-density polyethylene sheets with adhered bentonite granules. (See Chapter 12 for discussion of bentonite-based waterproofing products.)

Negative-side, cementitious coatings with iron or aluminum filings are fading in popularity. Crystalline/chemical coatings offer greater dependability, longer service life, and a self-healing property that enhances their long term waterproofing quality. Also called *chemical conversion* or *capillary* coatings, these solutions of organic and inorganic compounds react chemically with Portland cement and free lime in the presence of moisture. They fill capillaries and shrinkage cracks with long chain molecules crystallized within concrete. When shrinkage cracks occur after the concrete has cured, the reactivated crystallization process tends to fill the crack openings.

Synthetic materials are now used in place of natural materials even in ancillary components of waterproofing systems. Plastic drainage panels are now almost always used instead of pea gravel and granular backfill on plazas and against foundations. Geotextiles serve as filters for plastic drainage panels and granular backfill, and also as wrapping filters for footing drains. These synthetic materials offer greater filtration design control than bulkier natural materials. With their thinner cross sections, they also reduce excavation volume and costs.

Like improved filtering materials, waterstops have undergone several stages of improvement. Brittle bent metal strips were replaced by plastic and rubber-derivative dumbbells and bulb cross sections. Those were subsequently superseded by butyl and bentonite/butyl bars (see Appendix C). Along with greater durability, butyl and bentonite/butyl bars offer these advantages:

- simple installation (featuring lap joints);
- elimination of onsite welding;
- easy visual inspection of installation;
- stability during concrete casting operations; and
- complete encapsulation in concrete.

H. New standards for improved design

Waterproofing design has improved along with materials with an influx of industry standards. Plaza waterproofing design took a giant forward step with the Charles Parisie article[1] on waterproofing plazas that lead to the 1978 and 1983 publications of ASTM Standards C898 (for LAMs) and C981 (for built-up bituminous membranes).

These ASTM standards have revolutionized plaza design, through establishment of three vitally important design principles:

1. sloping decks for dependable drainage
2. use of thermal insulation
3. use of drainage courses above membranes

These standards prevented defects resulting from traditional practices that are unsuitable for modern waterproofing projects. For nearly a century, traditional waterproofing assemblies over setback terraces and sidewalk vaults consisted of four- or five-ply built-up membranes, without protection boards or thermal insulation. Membranes were covered with concrete or quarry tile in mortar setting beds. On terraces, membranes were installed on a one- or two-inch-thick concrete topping, underlain by cinder fill and sloped to drain. As an alternative design, decks were level with setting beds sloped to drain.

The omission of insulation from these traditional assemblies kept coal tar pitch membranes warmed by heat loss from the interior space, within a relatively narrow temperature range and thus reduced thermal contraction stresses experienced by membranes with insulation located on the warm side of the membrane.

The specific defect corrected by ASTM Standard C898 was the practice of sandwiching a waterproofing membrane between a structural concrete deck and a wearing slab. This defective detail propagated shrinkage cracks from the substrate below or movement from the topping above into the membrane (Figure 1.4). Traditional built-up bituminous membranes are especially vulnerable to splitting. (Their breaking strains are less than two percent at low temperature, compared with 200 percent or more for vulcanized rubber sheets and breaking strains ranging up to 100 percent for modified bitumens.)

The solution to this problem was isolation of the membrane from the paving with an intervening layer of insulation. Extruded polystyrene board has the properties required for waterproofing insulation. It has high compressive strength, required to resist traffic loads. It also has extremely low water absorptivity, required to prevent water buildup within the waterproofing assembly. Placed above a plaza waterproofing membrane, extruded polystyrene board reduces the membrane's temperature range,

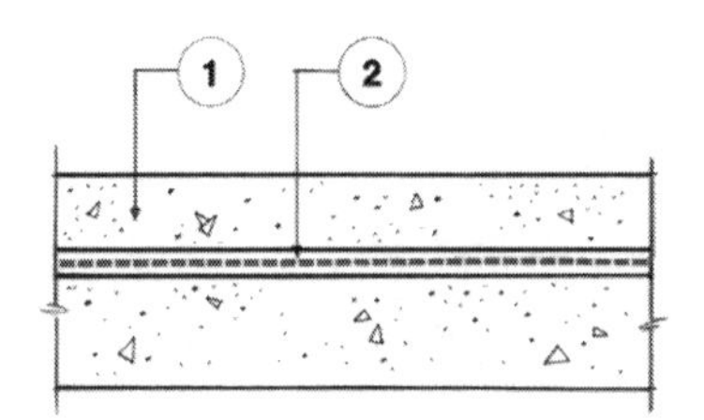

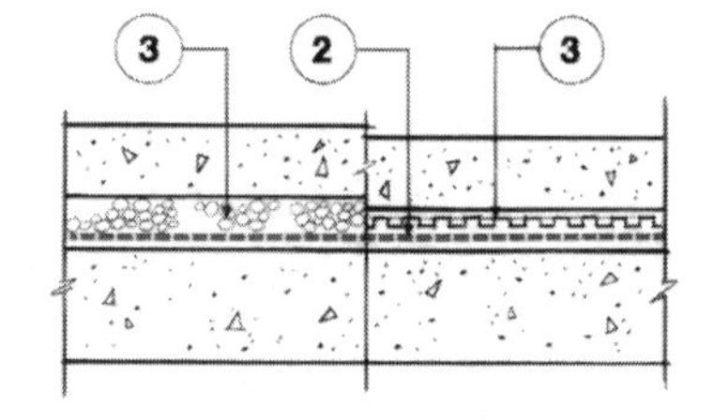

Figure 1.4 Traditional split-slab construction bonded the wearing course (1), to the membrane (2), as indicated on the left. Contemporary design divorces it with a layer (3) of gravel (left), or drainage composite (right)

thereby reducing the splitting risk from thermal contraction at low temperature. Drainage fills and composite panels normally used in conjunction with this foamed insulation, also help to isolate the membrane from the paving.

Membrane splitting from concrete cracking propagated upward from the deck is a less severe problem than movement propagated downward from the paving. The use of alternative adhered membrane materials – modified bitumens, EPDM, and PVC – with their much higher breaking strains, virtually eliminates the risk of membrane splitting from deck cracking. Even traditional built-up membranes are fairly safe from splitting at the narrow temperature ranges assured by locating insulation between cold winter air and the membrane.

Another design error corrected by the modern standards involved an increasing number of wearing-surface failures caused by improper drain location or improper drain installation. Drain elevation was set to carry water from a quarry tile wearing surface, with no drain for the membrane below. When drains with weep holes at the membrane elevation were installed to correct this latter defect, the weep holes were often clogged either with bitumen mopping or cement leached from the mortar setting bed. As a consequence, water was dammed in ever-widening circles around drains, causing efflorescence, popping tiles, disintegrating grout, and organic growth. These factors then combined to accelerate failure. Vegetation growth, for example, can ultimately result in roots penetrating a membrane.

Note

1 *ASTM Journal* "Architectural Considerations in Plaza Waterproofing Systems," BRI Fall Program 1971.

2 Principles of water management

Water management is essential to keeping a building dry. As with many aspects of building design, a gram of prevention is worth a kilogram of cure. Managing water flow to alleviate or eliminate leakage problems is better than just relying on waterproofing.

Good water management is analogous to good thermal design. Thermal design was inadequately managed in the ubiquitous glass walled-office towers built in the 1960s and early 1970s (right before the energy crisis and big increase in costs commanded designers' attention). Many architects ignored these buildings' tremendous solar heat loads, which ranged up to 10 times more through single-layered glass than with opaque insulated curtain walls. Previously, they just had their mechanical engineers design the air-conditioning system to handle these heavy loads. They ignored the additional capital and operating costs as well as the greater difficulty in assuring comfortable conditions in these heat-absorbing structures.

Just as wall-shading and heat-reflective glass reduce thermal loads, sloping below-grade structural slabs to drain can reduce hydrostatic loading of subsurface building components. Drainage panels or pervious granular backfills alleviate lateral pressure against walls. Granular fills under slabs-on-grade reduce capillary movement of moisture upward through voids in the subsoil, reducing or eliminating hydrostatic pressure.

These subsurface building components must also be designed to resist corrosive chemicals present in soil. Most products are usually resistant to chemicals with pH values of 3 to 12.

A. Hydrostatic pressure

Though structural design of waterproofed building components is beyond this manual's scope, a waterproofing designer needs to have at least an elementary understanding of hydrostatics. Below-grade structures must resist a combination of hydrostatic and soil pressures. These pressures generally range from 30 to 62.4 psf per foot of depth. They are obviously lower for dry, granular soils, which facilitate vertical subsurface water flow, and higher for wet soils, which approach purely liquid properties. Some soils – e.g. saturated clays and silts – exert lateral pressure (psf) equal to their densities (pcf), which can exceed that of purely hydrostatic pressure. This added earth pressure, however, affects only the structural design of a wall or slab. Despite the additional lateral compressive stress on a waterproofing membrane, the leakage threat to the waterproofing component comes from liquid water pressure alone.

Hydrostatic pressure increases linearly with depth, producing a triangular horizontal loading pattern (see Figure 2.1 and Table 2.1). At a depth of 10 feet, hydrostatic pressure is 62.4 × 10 = 624 psf, in all directions. The extreme pressures shown in Figure 2.1 assume a groundwater level at grade elevation, a conservative assumption that nonetheless indicates the scope of a waterproofing designer's problem.

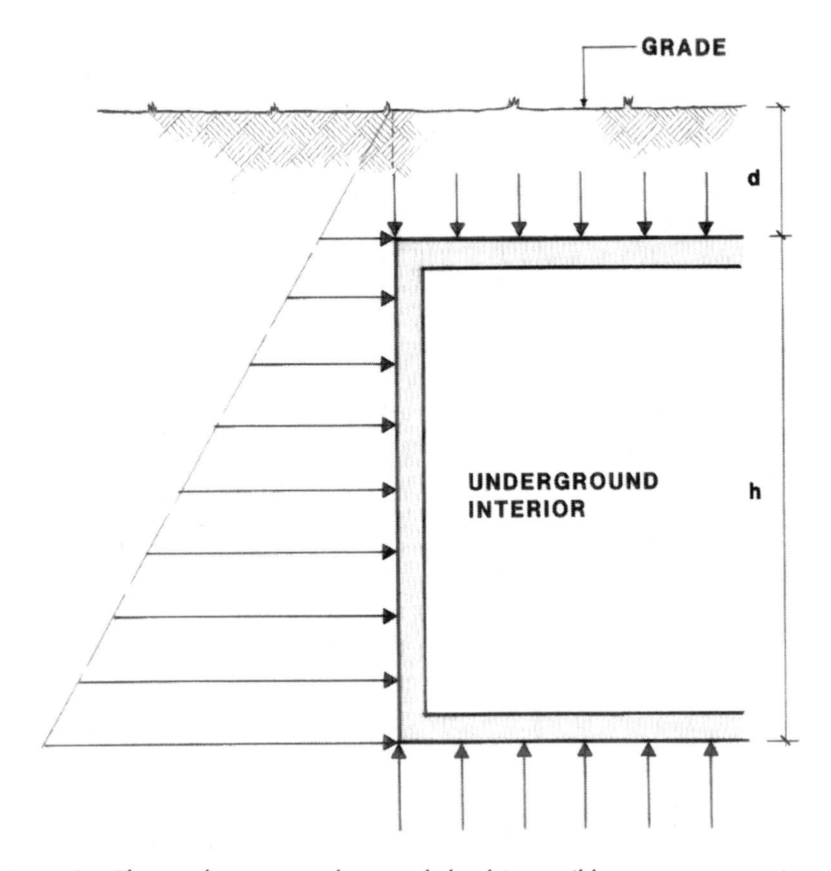

Figure 2.1 If groundwater extends to grade level (a possible temporary state in areas struck with torrential rain), hydrostatic pressure under the worst conditions can be as follows: p = wd on top slab; p = w(d+h) lateral pressure at the base of the wall, and p = w(d+h) upward pressure on the slab-on-ground. If d = 2 ft. and h = 10 ft., the lateral pressure at the base of the wall could equal 62.4 (2 + 10) = 749 psf

Table 2.1 Hydrostatic pressure increases with depth

Hydrostatic head feet	*Pressure per square inch pounds*	*Lifting pressure per square foot (under floor) pounds*	*Average pressure per square foot on wall surface affected pounds*
0.5	0.21	31.0	15.6
1.0	0.43	62.0	31.2
2.0	0.87	125.3	62.5
3.0	1.30	187.2	93.7
4.0	1.73	249.1	125.0
5.0	2.17	312.3	156.0
6.0	2.60	374.4	187.2
8.0	3.47	500.0	250.0
10.0	4.34	625.0	312.5
12.0	5.20	749.0	374.5
15.0	6.50	937.0	468.0
20.0	8.67	1248.4	624.2
25.0	10.84	1561.0	780.5
30.0	13.01	1873.4	937.0
40.0	17.34	2497.0	1248.5

Source: *Richey's Reference Handbook.*

Hydrostatic pressure on a building component occurs when the water table rises above it at any point. The water table (also known as the groundwater level), free-water elevation, and the phreatic surface, is defined by labs as the underground elevation where water is at atmospheric pressure. It may be at grade or hundreds of feet below grade. Hydrostatic pressure occurs on slabs-on-grade and foundation walls when they are below the groundwater line when surface water runoff temporarily raises the groundwater level, or when there is a perched water table. This is a condition where localized bodies of water are separated from the normal free-water elevation by impervious, unsaturated soils. Hydrostatic pressure occurs on structural slabs below grade when they are below the groundwater level, when flash floods occur, from heavy runoff from rising walls, and from clogged drains or undersized drainage systems.

Groundwater level, and consequent hydrostatic pressure can vary seasonally (generally highest in spring), daily, or even hourly. The resulting hydrostatic pressure may be intermittent, with groundwater level periodically rising above and falling below the waterproofed components' elevation; or it can be continuous when groundwater level remains constantly above waterproofed components. Spectacularly rapid changes in groundwater level can occur in semi-tropical locations like Florida, which sometimes experiences rainfalls of 10 inches or more in a day. Even in arid Arizona, flooding rains can suddenly raise water-table elevations in the clay desert soil. Under such conditions, the ground is swiftly saturated. Full hydrostatic pressure can be temporarily exerted against foundation walls, slabs-on-ground, and structural slabs until the surplus water drains away through the foundation soil.

Groundwater level is only the basic determinant of hydrostatic pressure. It also depends on the nature of the soil. Water rises by capillary action in most soils. This capillary rise varies from 11.5 feet in soils comprising microscopic particles (i.e., clay and silt) to zero in granular soils, where the large spaces between particles nullify capillarity. The top plane of the saturation zone – i.e., the space between groundwater level and the limit of capillary rise – lies between those extremes. Even in coarse sand, this saturation zone can extend hydrostatic pressure more than two feet above groundwater elevation (see Table 2.2).

Slabs-on-ground subjected to hydrostatic pressure must be designed to resist uplift, via one or more of the following methods:

- increasing slab weight (to counterbalance upward hydrostatic pressure)
- reinforcing the slab for flexural resistance and anchoring it to foundations or grade beams
- tying the slab to rock anchors

Table 2.2 Capillary rise

Capillary rise	*Soil type*	*Saturation zone*
11.5’	clay	5.7’
11.5’	silt	5.7’
7.5’	fine sand	4.5’
2.6’	course sand	2.2’
0.0’	gravel	0.0’

Note: Water rises in all soils except gravel via capillary action, with the results tabulated above. In clay and silt, the saturation zone extends nearly six feet above the groundwater table. Liquid moisture in the saturation zone exerts hydrostatic pressure that must be resisted by the waterproofed components.

Source: Data from *Construction Principles, Materials and Methods,* Second edn., *1972,* The Institute of Financial Education, Chicago, IL.

Passive resistance (prevention of hydrostatic pressure) can be accomplished by installing under-slab drains and footing drains directed to the storm-water system.

In addition to designing the system for the most severe hydrostatic pressures likely to occur, one must also consider other hazards when selecting a system. Swiftly flowing underground water can wash away bentonite, leaving the foundation wall defenseless. High hydrostatic pressure can force a waterproofing membrane into voids in its concrete substrate, causing shearing and flexural rupture. Hydrostatic pressure can also increase a membrane's water absorption, with consequent swelling. Membrane swelling can break a membrane's bond with its substrate, forming wrinkles vulnerable to flexural cracking and eventual leakage through local weak spots.

On negative-side waterproofing systems (see Chapter 7) hydrostatic pressure promotes water penetration into concrete walls and slabs through tie-rod holes, cold joints, and rock pockets. Other potential leak sources occur at buried conduit; tieback block-outs; and where soda cans, beer bottles, and other debris thrown into the formwork prior to casting operations create cavities in the concrete.

B. Site planning and landscaping

The most rudimentary solution to relieving water pressure in open areas with high water tables is to raise the building above it and use the excavated earth to grade away from it. Passive water and moisture control can be achieved via several methods, including:

- proper grading to control surface runoff. Grading should be sloped away from the foundation by at least five percent for the first 10 feet. The ICC code is more conservative, requiring a 1-in-12 (8.7 percent) slope.
- foundation planting, which reduces hydrostatic pressure by absorbing groundwater through plant roots.
- providing a layer of impervious soil or paving adjacent to foundations that diverts water flowing down the exterior wall to prevent it from creating hydrostatic pressure against the foundation.
- extending downspouts to discharge drain-water well away from foundations, which reduces the quantity of water that can percolate downward through the soil adjacent to foundation walls and slabs, where it can exert hydrostatic pressure.
- connecting downspouts to storm-water sewers. This is even more effective than extending downspouts away from the building as a means of relieving subsurface hydrostatic pressure. Unless the drainage system fails, this strategy removes all or nearly all surface water from the building perimeter.

- backfilling with pervious material and installing footing drains, which reduces hydrostatic pressure. In conjunction with previously listed remedies, porous backfill allows subsurface water to flow vertically to footing drains, which preferably discharge at some lower surface elevation. (See Figures 2.2 and 2.3.) Footing and under-slab drains draw down surrounding groundwater to relieve walls and floors from hydrostatic pressure. They also collect rainwater and snow-melt that percolates through the backfill.
- providing under-slab drainage blankets and drains.
- providing intercepting drains (uphill on hilly sites).
- draining impermeable surfaces around a hill-sided building (which can dam water).

C. Drain pipe materials

Drain pipe is manufactured in two basic types: rigid and flexible. Rigid pipe is either porous concrete or clay tile. Flexible pipe is made of corrugated polyethylene, PVC (corrugated and smooth-bore), bituminized fibers, or styrene rubber. Flexible pipe offers a major advantage over rigid pipe with its ability to accommodate (1) differential settlement without cracking under flexural stress and (2) silting up from soil penetration through the cracks.

Clay tile illustrates the liabilities of rigid pipe. It was the most prevalent drainage pipe material until the 1960s when its defects then became manifest.

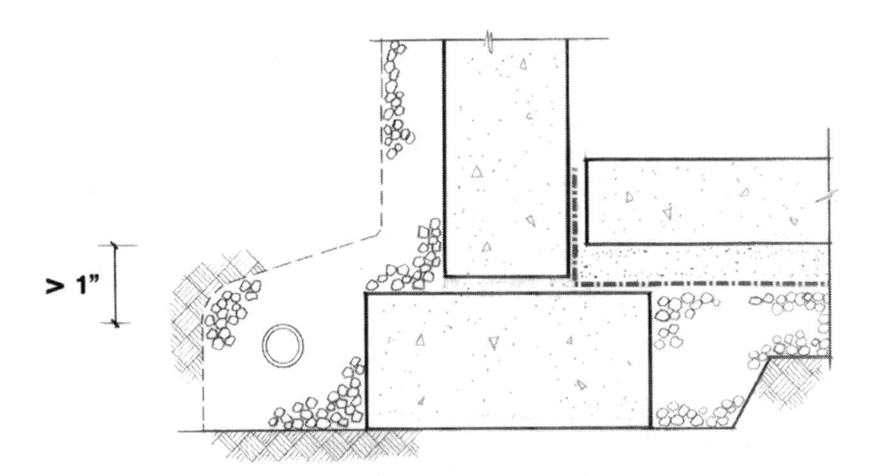

Figure 2.2 Detail of footing drain shows minimum aggregate blanket and minimum dimension from bottom of slab-on-ground to top of pipe. Pipe is usually four or six inches diameter and may be level, but a 0.4% to 1% slope is preferred

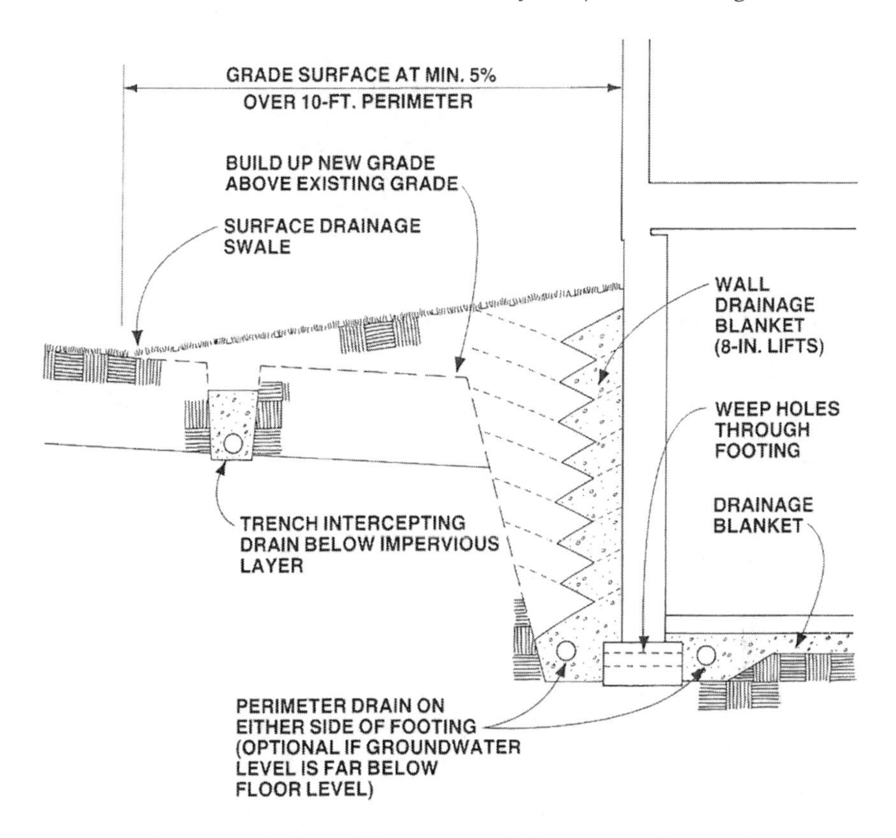

Figure 2.3 Cross section through basement wall depicts overall water-management strategy for relieving hydrostatic pressure against waterproofing or dampproofing membrane. (The University of Minnesota)

Its short, uncoupled sections were easily dislodged during backfill operations. Felt strips over open joints often failed and promoted silting, as did cracks in the brittle tile from flexural stress caused by settling and heaving.

Porous concrete, corrugated PVC, and corrugated polyethylene have become the most popular drainage materials due to their advantages. Despite its rigidity, and because of its superior flexural strength, porous concrete seldom cracks from differential settlement. It also has a unique advantage: allowing water infiltration around its entire perimeter. All other pipes are limited to infiltration through perforations and slots in the bottom half of their cross sections. Porous concrete is less vulnerable to silting than perforated or slotted pipes. Its microscopically sized openings permit less intrusion by silt particles than the much larger openings in alternative materials.

Corrugated pipe has a better balance of assets and liabilities than smooth-bore plastic pipe. Smooth-bore pipe offers reduced resistance to water flow, better self-flushing, and less deflection than corrugated pipe. Its reduced flexibility, however, makes smooth-bore pipe more vulnerable to cracking from differential settlement. Slotted, smooth-bore, corrugated PVC pipe (smooth-bore on its interior surface, corrugated on its exterior surface) combines the best properties of both types.

D. Basic drainage design

Drainage pipes are located at the toe of the footing or under the slab. At its highest elevation, the top of the pipe must be below the underside of the floor slab. At its lowest elevation, the bottom of the pipe must be above the bottom of the footing (see Figure 2.4). These rules prevent two deleterious effects:

1. Hydrostatic pressure underneath the slab.
2. Undermining of bearing soil by drain-water seeping below the soil-bearing elevation.

To assure compliance with these requirements, you must coordinate the drainage plan with the foundation plan. In some instances, the structural engineer may need to drop the footing elevation below the stratum determined satisfactory for bearing capacity.

On expansive soils, the footing may be cast on a drainage bed or a void form, with the pipe invert level with the bottom of the footing. On unstable soils, corrugated plastic tubing may be placed on top of the footing, which must then be at a lower elevation than the slab. Inboard placement of the footing is less effective than outboard, where it directly receives percolation from the grade.

Pipes should bear on a four-inch-minimum gravel bed surrounded with six inches of gravel. Gravel must be encased with a filter fabric. Positive drainage and scouring cannot be assured when drains are level. A pitch of one percent ($^{1}/_{10}$) is desirable, but can be reduced to 0.4–0.5 percent ($^{1}/_{17}$–$^{1}/_{20}$). Drainage systems under the slab should be connected to perimeter drains with two-inch-minimum weep tubes spaced eight feet apart through the footing.

Putting intercepting drains on hill sites prevents water from flowing against foundation walls and exerting hydrostatic (and even some minor hydrodynamic) pressure (see Figure 2.5).

Draining impermeable surfaces around a hill-sided building also relieves the foundation of subsurface hydrostatic pressure by diverting surface water harmlessly away from the building.

Draining water under slabs-on-ground by aggregate blankets and drainage fields routes accumulated water to daylight or drainage systems.

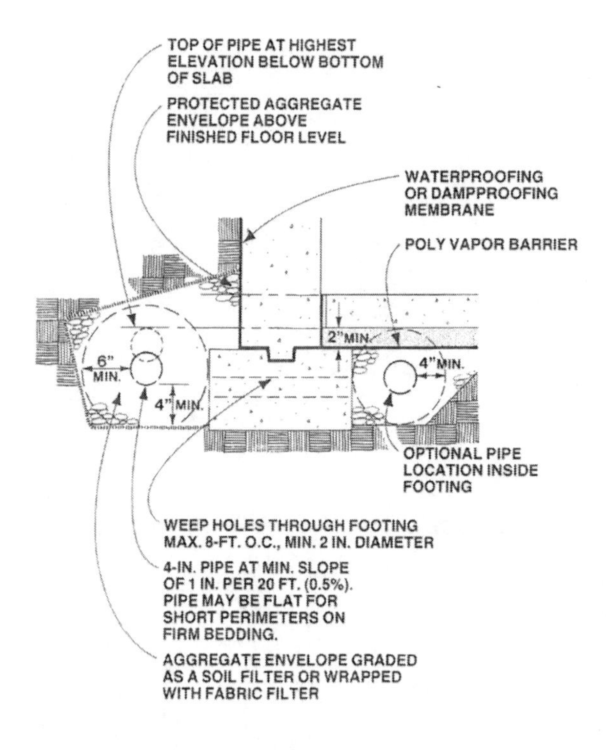

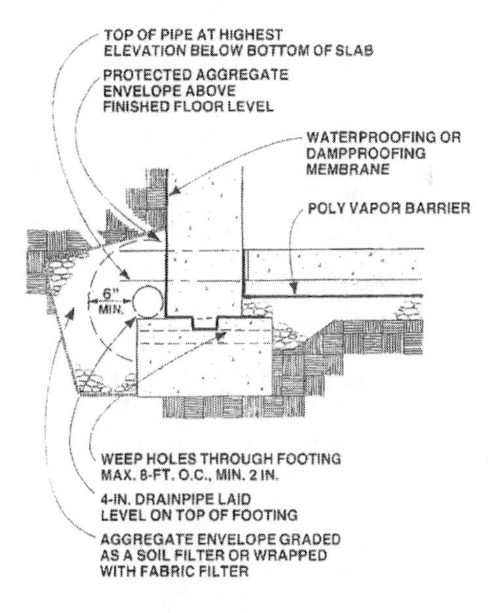

Figure 2.4 Wall footings with perimeter drains can locate drain pipes inside or outside footings. Inside location is less likely to clog drain, but requires much more disruptive repair operations if it does clog. (The University of Minnesota)

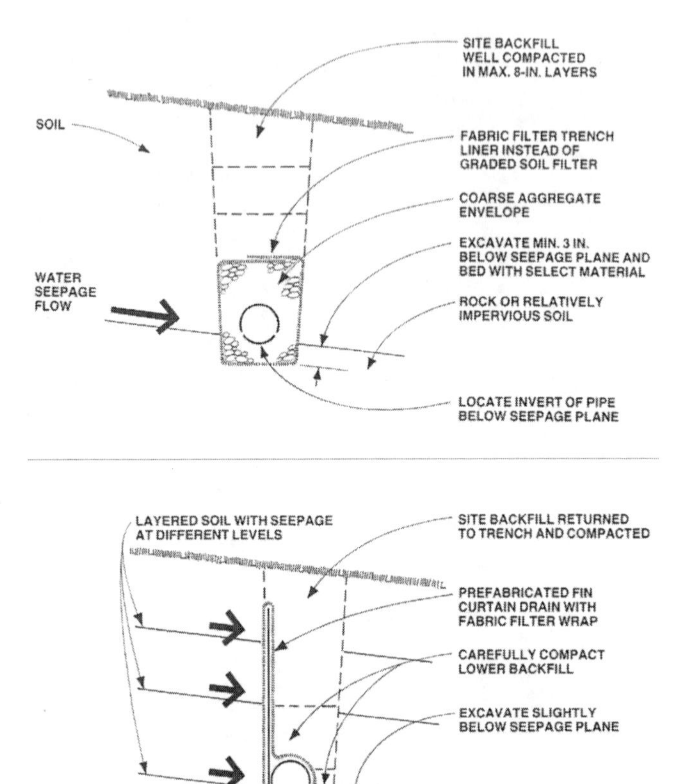

Figure 2.5 Intercepting drains featuring filter fabric (top) or a prefabricated fin curtain drain (bottom) can relieve hydrostatic pressure on sites where a waterproofed structure is downhill from subsurface water flow. (The University of Minnesota)

Where intermittent water from heavy rain storms or the occasional seasonal rise of the water table occurs, gravel or combined gravel and under-slab pipe drainage systems may suffice to reduce the frequency, duration, and intensity of hydrostatic pressure. Drainage backfills of gravel or sand, sometimes in concert with geocomposite panels, are also used to relieve intermittent pressure and groundwater accumulation.

Subsurface drains under slabs-on-ground, installed in a gravel drainage bed, can reduce the need to waterproof slabs exposed to intermittent water pressure. Design this system to extend the drains to daylight.

Depending on a pumping system to discharge the water can be risky unless back-up pumps and generators are provided. Moreover, many urban sewer authorities may not permit discharge of subsurface water into what may be an already overloaded storm drainage system. When the sewer system combines both storm water and sewage, adding subsurface drainage increases the load on a wastewater treatment plant. Also, subsurface drainage systems can draw down the water table and cause settlement of nearby buildings. If seasonally high water levels are anticipated, you and your client may accept the risk of installing an under-slab drainage system in lieu of waterproofing. The system can be either a thick drainage blanket or a thinner one with drain pipes in gravel-lined trenches. The thick blanket discharges water at a higher level than a piped system. The design of the gravel blanket thickness and pipe spacing is based on the permeability of the soil and drainage blanket and the height of the groundwater. Pipe sizes and spacing can be varied as dictated by cost.[1]

E. Subsurface drainage

Subsurface drains are a major factor in water management. Once water is induced to flow, it must be provided with a pervious material to permit it to drain freely for harmless disposal, usually via subsurface drains. Common drainage media include:

- drainage composites and insulating drainage panels, covered on one or both sides with geotextiles (filter fabrics described in section 2.F.)
- insulated drainage panels
- graded pea gravel
- coarse sand
- coarse gravel

Drainage composites and panels are inserted between the wall waterproofing membrane and backfill and on waterproofing over horizontal membranes. Marketed in a variety of forms, the most popular are three-dimensional sheets and fused plastic filaments laminated to geotextiles and other filter fabrics.[2] Boards or panels of grooved extruded polystyrene, coated expanded polystyrene and 6-pcf fiberglass are also used for drainage, although the latter two lack sufficient compressive strength for horizontal orientation. They are all covered on one or both faces with geotextiles. One manufacturer markets a combination insulation/drainage board with medium-density vertically oriented glass fiber as part of a liquid-applied membrane (LAM) system, which is limited to residential foundations.

Three-dimensional plastic drainage composites are an alternative to granular backfill. They are composed of three-dimensional drainage cores of plastic sheets or mats of fused filaments. Sheets are deformed and filaments interlaced to form a geomatrix that provides multi-directional water flow. In one product, the filaments are rectilinear. In the other, the filaments are non-directional. To reduce silting, both deformed plastic and filament cores are laminated to a filter fabric on one or both sides. These composite products range in thickness from 0.4 inches to 0.8 inches. Despite their higher material cost, the reduced labor cost for installing plastic drainage composites can offer lower total cost than using granular fill. They eliminate the need to install a separate filter fabric between the earth and granular fill since the filter fabric is laminated to the panel. Manufacturers also claim costs are reduced by eliminating protection board placed against the membrane. This may be false economy; membrane waterproofing requires prompt protection, and drainage panels are not always installed in a timely fashion. Further benefits of plastic drainage composites accrue from their thinner cross sections, which reduce the quantity of filter material. They also provide drainage at lagging and lot lines.

Different geotextiles are used for horizontal and vertical application. Take care to select the appropriate type. Composites are manufactured with a range of compressive strengths depending on the application. For horizontal surfaces, specify a high compressive strength of 9,000 psf. On foundations, a 6,000-psf panel is usually sufficient. Drainage composites have thinner cross-section dimensions. That makes them the preferred drainage medium on plazas that have design restrictions on thickness and where traditional two- or three-inch-thick sand and gravel would be unacceptable. However, you should anticipate a permanent initial deflection of approximately 10 percent under initial loading. Details should provide for this permanent deformation.

Plastic panels are spot-adhered to vertical waterproofing or laid loose on horizontal membranes. Panels function well on suspended slabs, but may distort and become disconnected from vertical surfaces when exposed to sunlight before backfilling.

Waterproofing on both horizontal and vertical surfaces should be covered with protection board before the panels are installed. This will prevent sharp filaments or panel edges from puncturing the waterproofing membrane. Improperly installed backfill may settle and dislodge the sheets.

Graded pea-gravel aggregate, also referred to as aggregate drainage blankets, consists of sand, gravel, crushed stone, blast-furnace slag, crushed shell, and coral. These aggregates are specified under ASTM D2487 for strength and stability. Note, however, that conformance to this specification does not ensure adequate drainage properties for backfill. Pea gravel refers to a naturally rounded stone falling between $^{3}/_{16}$ inch (sieve size No. 8) and $^{3}/_{8}$ inch roughly corresponding to ASTM C33, No. 8. Drainage aggregate is often required to provide a permeability of at least 25 times that of the surrounding soil.

Coarse sand is a finer form of aggregate drainage material, varying from roughly a No. 30 to a No. 8 sieve. In addition to its primary function of facilitating downward water flow adjacent to foundation walls, it also reduces vibration in seismic areas, harmlessly dissipating seismic forces via frictional forces between adjacent grains. Sharp sand is much more effective than rounded bank-run sand. Gap-graded sand (sand with nearly uniform grain size) accelerates the drainage rate. The smaller size, faster drainage rates, and limited availability often make it more expensive than pea gravel. (See Figure 2.6 and Table 2.3.)

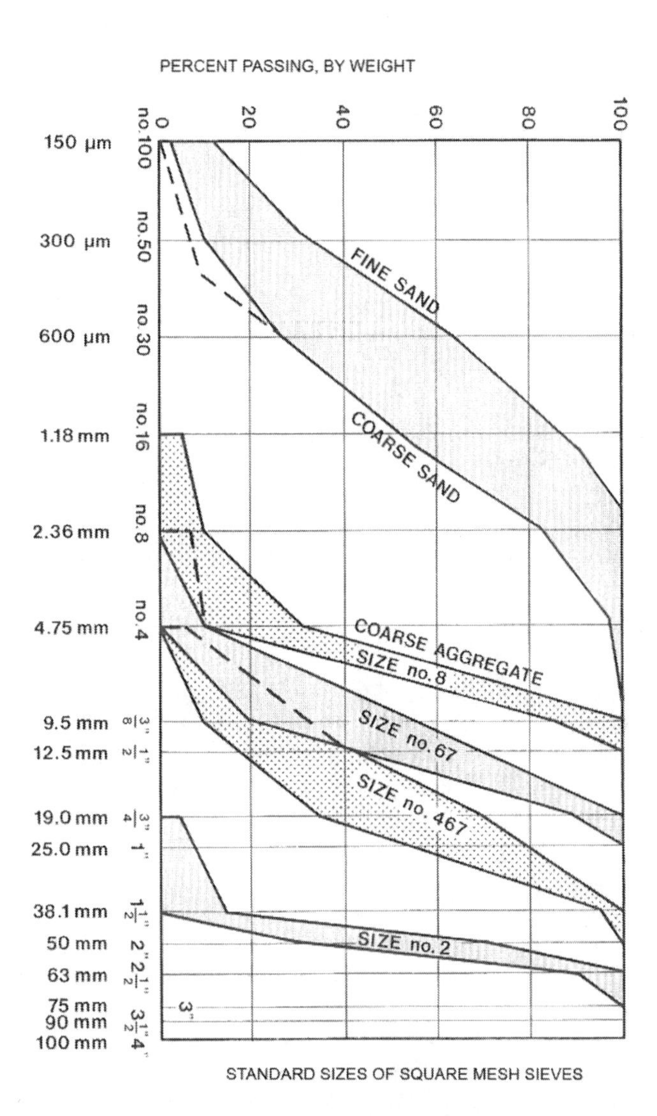

Figure 2.6 Distribution of soil grain sizes. Soils lacking a significant group of particle sizes between the largest and smallest are termed "gap-graded." Fairly uniform particle sizes are termed "open-graded." The permeability of soils is a function of the void area. Highly permeable soils have a small percentage of fines. (The University of Minnesota)

Table 2.3 Coefficient of permeability, K (cm/sec) is an index of vertical flow speeds of water flow through various types of soil. The major determinant of soil permeability is particle size: the larger the particles are, the more permeable the soil is. Large grain size creates large voids in horizontal cross sections through the soil, thus reducing frictional resistance to water flow through the soil. Uniformity of grain size also promotes vertical water flow, because heterogeneous mixtures of graded particles allow smaller particles to fill in voids between larger particles. As a consequence, the combination of size and percentage of fines is usually more important than the predominant particle size range in determining permeability

Soil type	*K=Coefficient of permeability (cm/sec)*
Clean fine to coarse gravel	10
Uniform fine gravel	5
Coarse sand (2mm)	0.4
Fine sand (.25 mm)	40×10^{-4}
Silt	0.05×10^{-4}
Clay	0.001×10^{-4}

Source: Simpson, Gumpertz & Heger, Inc.

Aggregate blankets and drainage fields under slabs-on-ground consist of aggregate placed along a wall, around a drain pipe, or under slabs to facilitate the flow of water from the backfill or subsurface grade into a drainage system for disposal above grade or into a sewer. They are composed of a six- to eight-inch-thick bed of 3/4-inch, single-size graded gravel, usually enclosed in a filter fabric. The size and gap grading prevent capillarity and retard silting. Pipe drains are installed in shallow trenches lined with aggregate and arranged to drain intermittent rises in the water table. If the subsoil contains excessive fines, consider installing a filter fabric under the blanket. Aggregate blankets do not relieve vapor transmission, so an underslab vapor retarder, or a suitable substitute, is still required.

F. Geotextile filter fabrics

Aggregate and coarse sand backfill against foundation walls were traditionally isolated from adjacent soil backfill by a layer of fine aggregate that impeded the infiltration of soil particles conveyed laterally by subsurface water flow. Filling the interstices between aggregate or coarse sand grains, fines impede the downward flow of water through the aggregate, with the harmful effect of increasing hydrostatic pressure against the walls.

In modern construction, fine aggregate filters are used instead of geotextile filter fabrics. With readily available manufactured filter fabrics as a better alternative, graded soil filters are difficult to justify. Filter fabrics, with their finite permeability and soil-retention qualities, allow you to precisely match a fabric to specific drainage requirements. Moreover, fabrics are lightweight, easily handled, simply installed, and readily visible for quality control.

Geotextiles are fabricated from polypropylene, polyester, or nylon fabrics; either woven or non-woven (spun-bonded), and designed for a narrow range of permeability. A universally accepted guideline or standard for selecting filter fabrics has not been established. Some manufacturers determine the coefficient of permeability in cm/sec. (usually .02 to .08, "moderately slow") by measuring the rate of water transfer through the fabric at some constant head. Others measure permeance in gal/min/sq ft for a lower constant head. The U.S. Army Corps of Engineers rates filters by the Equivalent Opening Size (EOS), approximately the sieve size closest to the filter fabrics' openings. The maximum EOS is 70 (.0083 inches), which is not greater than 10 percent open area; the minimum is 100 (.0059 inches), which is not smaller than four percent open area.

Filter fabrics laminated to plastic drainage composites are used to encapsulate porous backfill and pipe drains. None of the fabrics are designed to withstand prolonged exposure to ultraviolet radiation.

G. Water management system failures

The most common failures of aggregate drainage systems are infiltration of fines clogging percolating aggregate and roots clogging drainage pipes. Clogged drains located on the outside of the footings are more common but are more easily cleared by root clearing snakes and flushing. Drains under slabs inside the footings are less likely to clog from either fines or roots. They are, however, much more difficult to clear, because of their relative inaccessibility.

The most likely causes of filter-cloth failure are (1) a failure to lap the cloth sufficiently to prevent infiltration of fines, or (2) cloth tearing during backfilling operations.

Drainage composites may be dragged down when backfill is improperly placed or consolidated. When protection board has been omitted, the edges of some products can cut into or displace the membrane.

H. Checklist

1. Consider exploiting the building site's potential for alleviating waterproofing problems by raising the building and using excavated soil to grade the site to promote surface runoff.
2. Check the building code for the minimum grade slope to promote runoff. If the building code requirement is less stringent than five percent, use this slope as a minimum.
3. Coordinate the drainage plan with the foundation plan, to assure proper elevation of drainage pipes.

4. Slope below-grade structures to drain, with a minimum one percent slope.
5. Check the soil for chemical contaminants that can attack waterproofing materials.
6. Discharge downspouts into a storm sewer or dry well located some distance away from the foundation.
7. Consider aggregate blankets under slabs-on-ground for hillside sites.
8. Specify gap-graded sand and gravel to increase porosity.

Notes

1 The design is beyond the scope of this manual. For further information see Cedergen, H.R. 1962, "Seepage Requirements of Filters and Pervious Bases" Paper 3363, ASCE Transactions, vol. 127, part 1, pp 1099–1113 and 1967 "Seepage, Drainage and Flow Net" New York, John Wiley & Sons.

2 See ASTM D7001 "Standard Specification for Geocomposites for Pavement Edge Drains and Other High-Flow Applications".

3 Dampproofing

A. Distinguishing dampproofing from waterproofing

Waterproofing resists hydrostatic (and even minor hydrodynamic) pressures that can reach considerable magnitude: 1,000 psf for walls or slabs 16 feet below the water table. Dampproofing merely resists the flow of water vapor through a building component. ASTM defines dampproofing as the treatment of a surface or structure to block the passage of water in the absence of hydrostatic pressure. In usage, positive side membrane waterproofing resists both vapor migration and hydrostatic pressure from liquid moisture. Dampproofing just resists vapor migration.

Regardless of construction material – cast-in-place reinforced concrete; concrete; or brick masonry – dampproofing or waterproofing is required for slabs-on-ground and foundation retaining walls. Though well-constructed reinforced concrete is inherently waterproof, concrete inevitably cracks due to:

- plastic or drying shrinkage;
- live-load deflection resulting from traffic;
- mechanical vibration;
- seismic movement;
- building settlement; or
- frost damage.

Water can penetrate tie-rod holes, "cold" (i.e., construction) joints, expansion joints, and through inadequately vibrated, honeycombed voids within concrete. For these reasons, underground structures require either dampproofing or waterproofing. As explained in Chapter 5 "Waterproofing," elevated structural slabs over underground spaces always require waterproofing. This requirement holds for both structural slabs below and above grade, either earth-covered or with separate wearing surfaces (e.g., plazas, terraces, or promenades). These slabs require waterproofing because of the constant threat of liquid moisture above them exerting downward hydrostatic pressure.

Other components requiring dampproofing or waterproofing include planters and retaining walls, where water penetration might stain or deteriorate exterior finishes.

B. Preliminary investigation

The preliminary steps in deciding between waterproofing and dampproofing are consulting the building code and conducting a site survey. Building-code requirements may pre-empt your choice with mandatory minimum requirements that take legal precedence over your judgment.

Dampproofing and waterproofing are covered under Section 1807 in the ICC International Building Code 2015 as of this writing. It requires waterproofing or dampproofing where walls or portions thereof retain earth and enclose interior spaces and floors below grade. Codes require:

- a subsoil investigation to determine whether hydrostatic pressure is present or will exist during intermittent groundwater
- dampproofing where groundwater table is lowered and maintained at an elevation at least six inches below the bottom of the slab-on-ground
- perimeter drains where walls are dampproofed

The relevant section of ICC code is **1806.1.3 Groundwater control**, which reads, "Where the groundwater table is lowered and maintained at an elevation not less than six inches below the bottom of the lowest floor, the floor and walls shall be dampproofed in accordance with Section 1806.2. The design of the system to lower the groundwater table shall be based upon accepted principles of engineering which shall consider, but not necessarily be limited to: permeability of the soil; rate at which water enters the drainage system; rated capacity of pumps; head against which pumps are to pump; and the rated capacity of the disposal area of the system."

A site survey should focus on determining the presence of the following hazards:

- water levels (as an index of hydrostatic pressure)
- flowing underground water
- corrosive soil chemicals or petroleum pollutants

The presence of any of them eliminates dampproofing as an alternative to waterproofing.

After completing these two steps, consider risks, costs, and benefits before making a decision. Additional construction work needed to qualify a project for dampproofing will raise project costs. Drainage piping must be increased in size and complexity. Pumps and backup pumps, with standby generators, must be a cost-effective tradeoff for eliminating the risks and costs of waterproofing. The municipality must permit drainage into its storm-sewer system, which may already be operating at capacity, without excessive charges.

C. Water-vapor migration

Diffusion of water vapor is roughly analogous to thermal conduction in heat-energy transfer. Whenever there is a difference in vapor pressure on opposite sides of a building component, vapor naturally tends to flow from the higher toward the lower pressure side. This pressure-equalizing phenomenon is a consequence of the kinetic theory of gases, according to which the random motion of molecules in a mixture tends to disperse these molecules in equal proportions throughout the mixture. In the atmosphere, water vapor is only one of several gases – predominantly nitrogen and oxygen – but also including argon, carbon dioxide, and other gases.

Vapor pressure depends on two variables: temperature and relative humidity (RH). As shown in the curves of Figure 3.1, vapor pressure increases exponentially with increasing temperature. The inevitable presence of liquid moisture in soil, sand, and aggregate backfill will maintain its

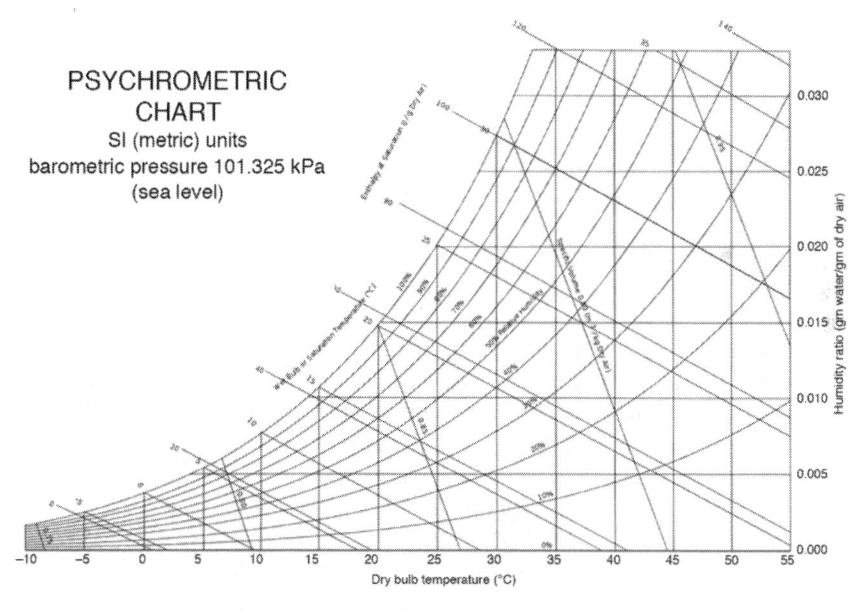

Figure 3.1 This chart plots vapor pressure for a given temperature and relative humidity (RH). Underground vapor pressure represented by the horizontal scale, can be assumed to be the intersection of the 100% RH curve with the temperature (vertical scale). Interior vapor pressure will vary much more than underground vapor pressure, because of virtually unchanging underground temperature

RH at or close to 100 percent whereas the RH of the underground interior space will normally be substantially less than 100 percent. Vapor flow will thus be impelled toward the interior. It moves laterally from the back behind a retaining wall or vertically (upward or downward) through a slab-on-ground below or a structural slab above the underground space unless the underground space is at a significantly higher temperature than the surrounding soil. (At 76°F, 50 percent RH, vapor pressure roughly equals the vapor pressure at 55°F, 100 percent RH.) Earth temperatures vary much less than air temperatures, because of the insulating value and heat capacity of the soil. Soil temperature near the surface varies much more than temperatures at greater depths, which even in climates with occasional subzero temperatures, remain close to the freezing level. Thus, the vapor-pressure gradient will normally be directed from the outside toward the inside. Moreover, when the underground space must be maintained at relatively low RH – such as 40 percent – dampproofing, at a minimum, becomes mandatory.

Water vapor flow through a building component poses a threat of condensation either within that component or within the interior space when it contacts a surface at the dew-point temperature (i.e., the temperature representing 100 percent RH).

Condensation is a greater threat in underground spaces for two reasons:

1. Unconditioned underground spaces tend to be cooler in summer because of the insulating effect of surrounding backfill and foundation soil.
2. Air circulation, which helps to disperse water vapor and prevent condensation, is generally less efficient in underground spaces, especially in non-air-conditioned spaces, because there are no windows or joints for air to leak through and thus convey water vapor out of the space.

Seasonal temperature changes tend to make dampproofing mandatory for interiors with humidity limitations. Since the soil RH must be conservatively assumed at 100 percent, summer soil temperatures around 70°F for soil near the surface create a strong vapor drive toward an interior space maintained at 72°F, 30 percent RH (see Figure 3.1). Even though vapor drive normally reverses during the winter in most continental US locations, this situation nonetheless calls for dampproofing. This demand for dampproofing during summer conditions makes vapor control a more demanding problem for a waterproofing designer than for a designer of low-slope roof systems. In roof design, the vapor problem is normally reversed, with vapor flow outward from the interior into the roof system rather than inward toward the occupied space. During spring and summer, solar heat and higher summer vapor pressures reverse vapor-flow direction. Liquid moisture, accumulated within the roof assembly, evaporates and vents into the building interior. This "self-drying" roof mechanism has been rigorously investigated by Tobiasson and others.[1] There is no analogous "self-drying" mechanism for underground spaces.

You must limit vapor flow in the same direction (exterior-to-interior) throughout the year. A reversal of vapor flow during the summer merely alleviates this problem. It does not solve it.

D. Dampproofing or waterproofing?

Deciding between waterproofing and dampproofing is simple. As stated in Section A, the only components for which you can consider dampproofing are foundations walls and slabs-on-ground. Structurally supported slabs inevitably run the risk of hydrostatic pressure and thus require waterproofing. To determine whether foundation walls and slabs-on-ground require only dampproofing you can consult the flow chart depicted in Figure 5.2.

In Figure 5.2, the key question, "Is a slight leak risk acceptable?" can be stated more technically: Is there any threat of hydrostatic pressure in addition to vapor pressure? If so, there is at least some minimal risk of water leakage. In an underground garage, you might tolerate a slight leak risk. In an auditorium, however, you could not tolerate *any* risk. For such occupancy, you could not tolerate any dampness on the interior face of the exterior walls.

As a general rule, dampproofing is required on all subgrade foundations and slabs-on-ground not subject to continuous, or even intermittent, hydrostatic pressure (in which event they require waterproofing, see Chapter 2 "Principles of Water Management"). Whenever melting snow or heavy rainfall may occur in poorly drained soils or on flood plains, foundation walls should be waterproofed where interior dampness cannot be tolerated. Slabs-on-ground under gymnasiums or other uses with water-sensitive flooring similarly require waterproofing, except under subsequently discussed conditions.

If you eliminate the need for waterproofing, you can determine if you can also eliminate dampproofing. You may be able to if the

- site has exceedingly dry soil;
- bottom of a slab-on-ground is above an adjacent grade; or
- applicable building codes do not require dampproofing as a minimum moisture protection. Remember that waterproofing is required if the groundwater level cannot be maintained at least six inches below the bottom of the slab-on-ground.

Dampproofing is applied to the following components:

- slabs-on-ground
- foundation walls

- retaining walls with finish bonded to wall (but not for concrete walls with cavity and brick facing)

E. Dampproofing walls

Dampproofing is required for foundation and retaining walls, with one notable exception: foundation walls with a vertical cavity through which water can flow down and discharge through weep holes (see Figure 3.2).

Dampproofing of foundation and retaining walls is always applied to the exterior (wet) face. Planters are similarly dampproofed on the wet (interior) face (see Figure 3.3). Dampproofing of the inside (dry) face of retaining or foundation walls is banned because of potential blistering from vapor pressure exerted in small chambers formed between the inside wall face and the dampproofing membrane. (Negative-side – i.e., dry-side – waterproofing requires some vapor permeability to alleviate the hazards of blistering; see Chapter 7.)

The most common dampproofing material for walls is a bituminous coating, either solvent-based (cutback asphalt) or emulsion. It is brushed, sprayed, roller-coated, or troweled onto the substrate. It may be liquid or mastic, reinforced with organic or inorganic (glass) fibers. Other dampproofing coatings include rubber polymers, rubberized asphalt, and acrylic.

Emulsions have several disadvantages. Being water-based, they cannot be

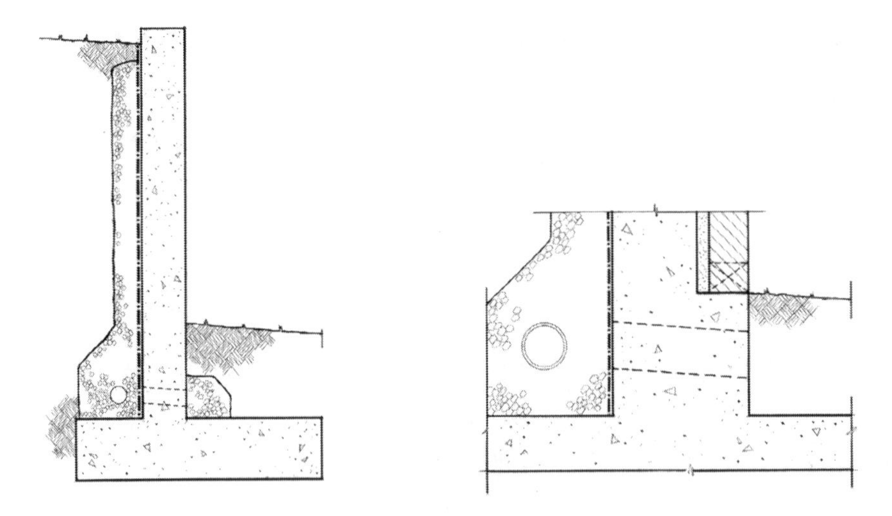

Figure 3.2 Retaining walls are dampproofed to prevent efflorescence on the exposed face. Brick facing is detailed on insert (right). Pervious fill and a collector drain are mandatory

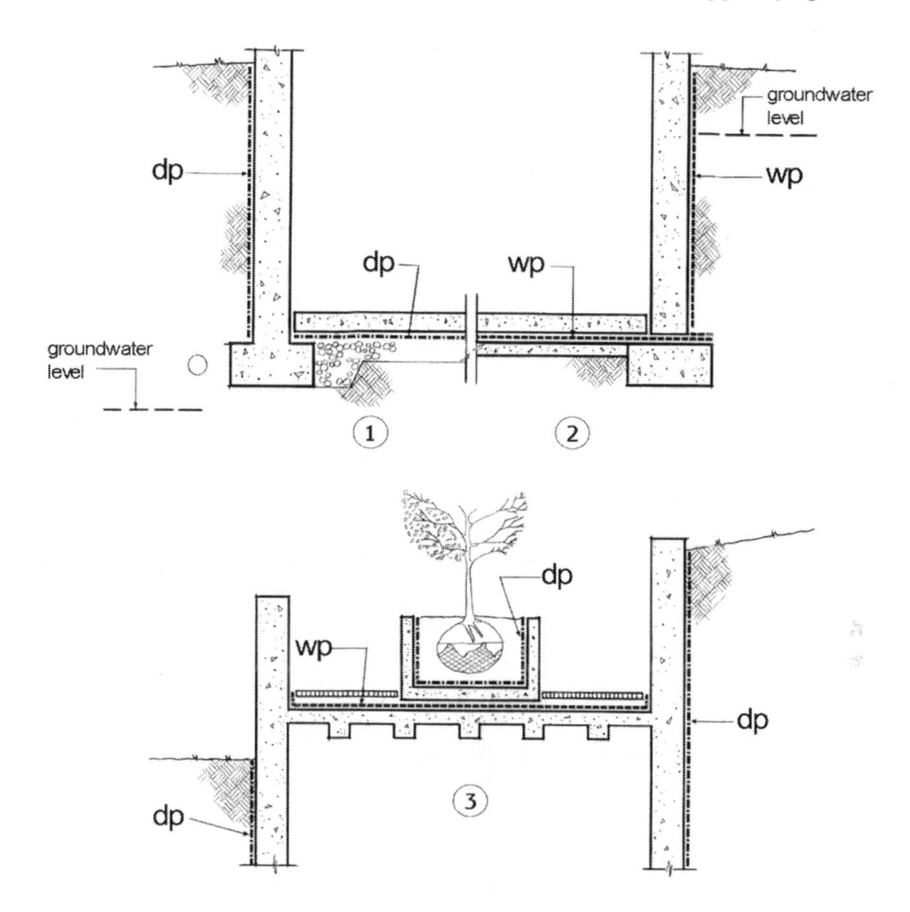

Figure 3.3 Cross-section (1) shows condition permitting dampproofing of foundation and slab-on-ground, when highest groundwater level is at least six inches below the bottom of the slab-on-ground. When the groundwater level is above the slab-on-ground, waterproofing is required for both foundation wall and slab-on-ground, as shown in (2). Even where the groundwater level is below the slab-on-ground, waterproofing is required for a structural plaza slab and a planter supported on it, as shown in (3) above

stored or applied at freezing temperatures. If not fully set before backfilling is completed, the carrier water may not escape. This prevents formation of a uniform film and thus produces an ineffective vapor retarder.

Dampproofing membranes often use the same materials and application techniques as waterproofing membranes. The dividing line between dampproofing and waterproofing has been set at 10-mil (0.25-mm) film thickness by Maslow. The thicker the coating, the longer the service life. In thin coatings, essential oils, which prevent embrittlement of the coating, may hydrolyze or migrate into the backfill soil.

Masonry foundation walls require a 3/4 to one-inch-thick cement parge coat, prior to application of the dampproofing, according to most authorities. It provides a smooth substrate instead of the rough masonry surface. It also seals mortar joints and adds another water barrier to the foundation wall.

F. Dampproofing slabs-on-ground

Slabs-on-ground require dampproofing or waterproofing except in a few instances – e.g., in arid regions with no soil or drainage problems and no irrigation (unless the governing building-code requires treatment.) ACI 302.IR-89 also suggests that you can omit a vapor retarder on well drained home sites, subject to these additional conditions:

1. the water table is perpetually below the ground surface;
2. a freely draining substratum of coarse granular fill is installed; and
3. a floor covering unaffected by moisture is installed.

Slabs-on-ground can be dampproofed either above or below. The most common, and generally better option is below the slab. Plastic film vapor retarders are the most common below-slab dampproofing (see Figure 3.4). The above-slab location is better in spaces where moisture control is most critical: zero tolerance situations. Gymnasium and dance floors are the most common. Wood flooring spans sleepers separating it from the concrete slab's top surface (see Figure 3.5). Note, however, that this top-of-slab dampproofing is acceptable only under the following conditions:

- the water table is perpetually (i.e., normally and intermittently) at least 12 inches[2] below the slab
- footing drains (or footing drains plus under drains) are installed

Unless these two conditions are satisfied, under-slab waterproofing is required for slabs-on-ground topped with wood floors or other moisture sensitive finishes.

Partitions and columns pose another limitation on top-surface dampproofing of slabs-on-ground. They interrupt the vapor retarder and thus violate the dampproofing's integrity (see Figure 3.6). In such instances, under-slab dampproofing or waterproofing is a safer alternative.

Substituting a vapor retarder for waterproofing normally entails risks. A vapor retarder depends on the integrity of the film and its seams. Field conditions for a vapor retarder for slab-on-ground normally make it more difficult to maintain field-seam integrity than for a wall or roof system.

Wood and some vinyl floor coverings and their water-sensitive adhesives make vapor retarders especially risky. Under these conditions, the additional expense of heavy-duty panels or waterproofing is well worth the price to avoid the expense of replacing a moisture-damaged floor covering.

Figure 3.4 A plastic vapor retarder is installed over a tamped gravel substrate and the seams are taped as shown. Note that the vapor retarder is sealed to the foundation wall to provide continuity. Brick chairs supporting the reinforcing are preferred to wire chairs to minimize penetration of the sheet

Sleepers for a wood floor covering on a slab-on-ground with top-surface dampproofing should be attached to the slab with water resistant adhesives rather than mechanical anchors. Nails and other mechanical anchors puncture a vapor retarder located on a slab's top surface.

When the top of the slab is dampproofed, a substratum of granular material should be placed below the slab, with aggregate properly graded to prevent moisture rise (and consequent hydrostatic pressure) via capillary action.

If you are committed to top-surface dampproofing, do not use the belt-and-suspenders' approach of adding another vapor retarder below the slab. Vapor pressure from residual moisture in the concrete slab, denied any means of venting via either the top or bottom slab surface, can disbond the top vapor retarder and ultimately rupture it.

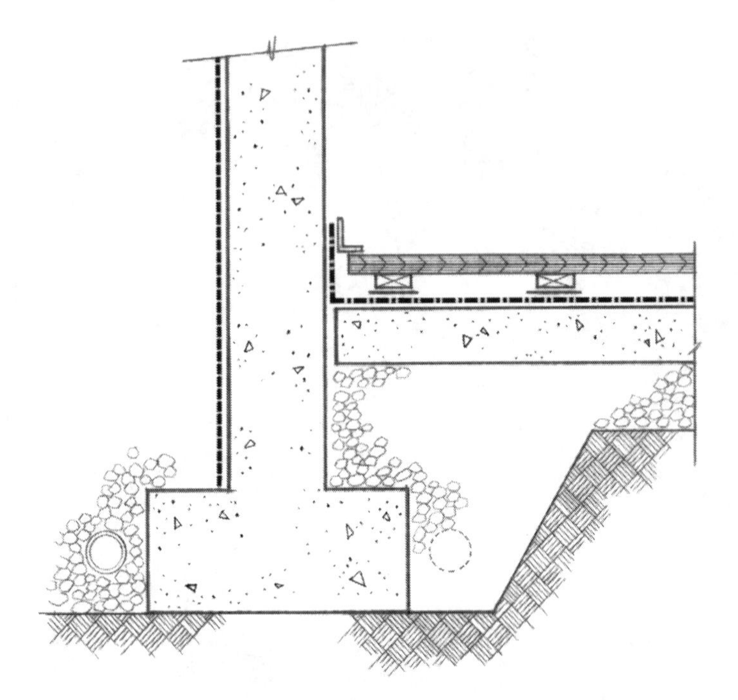

Figure 3.5 When slabs-on-ground are covered with wood gymnasium or dance floors, where wood flooring spans sleepers separating it from the slab, dampproofing can sometimes be applied to the slab's surface, but only under certain subsoil conditions. The air space intercepted between slab and flooring should be vented at junctions with the wall at the expansion joint, as shown in the above detail. This is an additional precaution to avoid moisture damage to the floor from water vapor migration penetrating the vapor retarder. This maintains an equal relative humidity below and above the wood floor

Many floor finishes and coatings applied to concrete fail because they are moisture sensitive. Those that can resist moisture may be installed with moisture sensitive adhesives. The latter has accounted for many recent failures where solvent-based adhesives have replaced water-based adhesives to conform to limitations on VOCs (volatile organic content) dictated by the Clean Air Act.

Many liquid-applied coatings (membranes) are sensitive to moisture in concrete slabs. High levels of moisture can exist in the following conditions:

- slabs cast on grade without functioning under-slab vapor retarders
- suspended slabs cast over non-vented steel formwork
- suspended slabs over occupancies with a high relative humidity, such as swimming pools and commercial kitchens

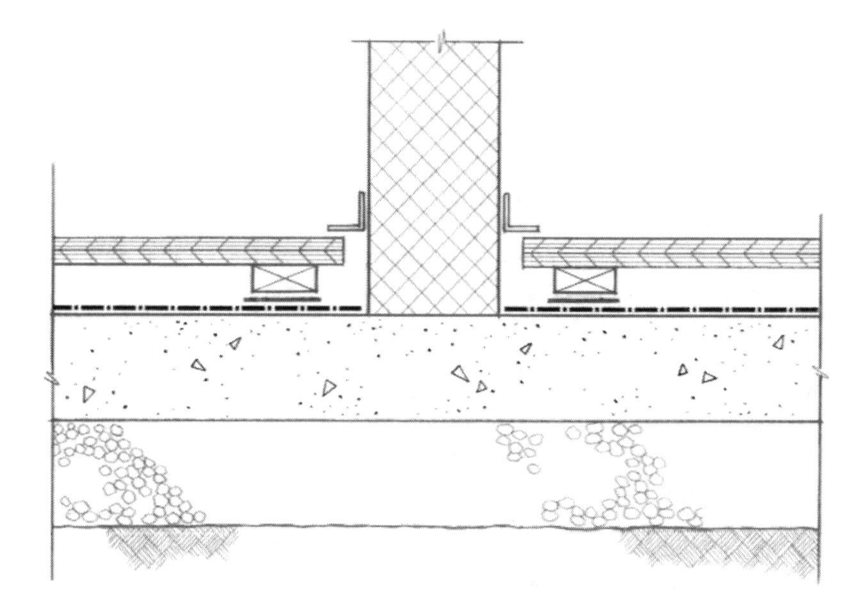

Figure 3.6 Partitions or columns can interrupt a vapor retarder on top of a concrete slab-on-ground. Under these conditions, a below-slab location for the vapor retarder is preferable. Waterproofing will usually be required.

- suspended slabs over unvented crawl spaces
- slabs with lightweight aggregates
- slabs with w/c ratios in excess of 055
- slabs that have cured less than 90 days

Manufactures of floor finishes and coatings or membranes require that the concrete slab be tested to ensure that the moisture is within sustainable limits and will not result in failure. ASTM tests for determining the level of moisture in concrete are listed in Appendix D. ASTM F1869, the calcium chloride test, is used to determine the MVER (Moisture Vapor Emission Rate), which is measured in pounds per thousand square feet in 24 hours. The generally accepted maximum can range from three to five pounds.

ASTM F2170 involves placing probes in the concrete and taking readings with a hygrometer. It is used to determine the internal relative humidity, which is established in ASTM F710 as 75 percent. Generally, achieving this requires one month of drying time per inch of concrete. Install vapor retarder parallel with the direction of concrete pour. ASTM E1643 ("Standard Practice for Selection, Design, Installation, and Inspection of Water Vapor Retarders Used in Contact with Earth or

Granular Fill Under Concrete Slabs") requirements apply to reinforcement supports.

Both methods require 72 hours before testing can begin, but the internal relative humidity test is regarded as superior because it measures the moisture deeper within the slab than the calcium chloride test does. The plastic sheet method (see ASTM D4263) is the most widely used, but the least reliable. It consists of taping a plastic sheet to the concrete and observing moisture that may have accumulated after 18 to 24 hours. The results are for the measured period, which may not be representative.

Installed *before* concrete casting, under-slab vapor retarders are tougher than above-slab vapor retarders. They are available in polyethylene sheets, asphalt/polyethylene composite sheets, coated granular-surfaced roofing felts, and polyolefin or polymer-modified bitumen sheets. Polyethylene should be high density (HDPE), which is relatively inert and resistant to many different chemicals. Polyethylene sheets are marketed in clear and black, reinforced and unreinforced. One proprietary sheet consists of two sheets of kraft paper laminated with asphalt, reinforced with glass fibers and covered on both sides with polyethylene. Another consists of an asphalt core board with a PVC sheet bonded to coated asphalt felts.

Under-slab vapor retarders should be selected for their low permeance, and resistance to tearing, puncture, and degradation from soil chemicals and bacteria. ASTM C755 recommends a perm rating below 1.0, but, in fact, most vapor retarders on the market have permeance well below 0.5 and a permeance of even 0.1 perm is not unusual. Polyethylene is commonly specified in 6- to 10-mil thickness. Use of two 6-mil sheets, to reduce puncturing and provide seam redundancy, is a prudent practice. However, use of polyolefin sheets scored higher than polyethylene in ASTM E1745. A 2-mil polyamide scored better than a 6-mil polyethylene over time.

G. Slab-on-ground installation details

Installation of vapor retarders under slabs-on-ground should follow recommendations in ACI Document 302.IR-89 and ASTM E1643 Standard Practice for Installation of Water Vapor Retarders Used in Contact with Soil or Granular Fill Under Concrete Slabs. These documents recommend a substratum of ³/₄-inch, single size, graded gravel under the vapor retarder. The gravel prevents capillarity.

At the turn of this century some designers advocated installing a three-inch-thick layer of sand over the vapor retarder. Its perceived function was to permit the slab to dry downward and reduce curling. It would also protect the vapor retarder from traffic and puncture from reinforcing chairs. Arguments for and against this are discussed in the Appendices of ASTM E1643. The majority of experts believe that it is unnecessary because satisfactory slab finishing operations can be obtained with water-reducing additives, and the potential for a rainstorm flooding the sand cushion, as well as the displacement from traffic, counteract its advantages.

Footing drains are always required when the vapor retarder is below an adjacent grade. Footing drains *and* under-slab drainage are required when the vapor retarder is below the adjacent grade and the highest intermittent groundwater elevation is within six inches of the bottom of the slab. The subsurface drains should be conducted to daylight for gravity flow, or connected to a pump. When the bottom of the footing is located below the groundwater level, to prevent "rising damp" (the capillary rise of water in the wall from the normally wet footing), the vapor retarder should be extended across the footing, as shown in Figure 3.7. Where the top of the wall footing is more than a foot below the bottom of the slab-on-ground, an alternative method is shown in Figure 3.8, with the vapor retarder returned to the top of the slab-on-ground.

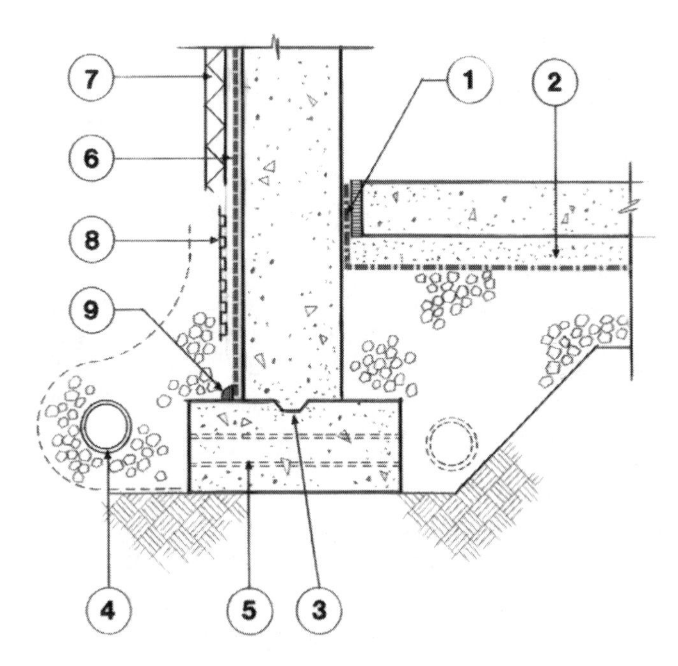

Figure 3.7 In this detail, with the top of the wall footing at least one foot below the bottom of the slab-on-ground, the under-slab vapor retarder is returned to the top of the slab (1). Other components in this detail are: (2) sand fill; (3) key; (4) footing drain four- to six-inches in diameter; (5) 2-inch-diameter drains through footing; (6) waterproofing membrane with protection board, or; (7) low-density insulation as protection layer; (8) drainage composite or gravel drainage fill; (9) liquid component of waterproofing system

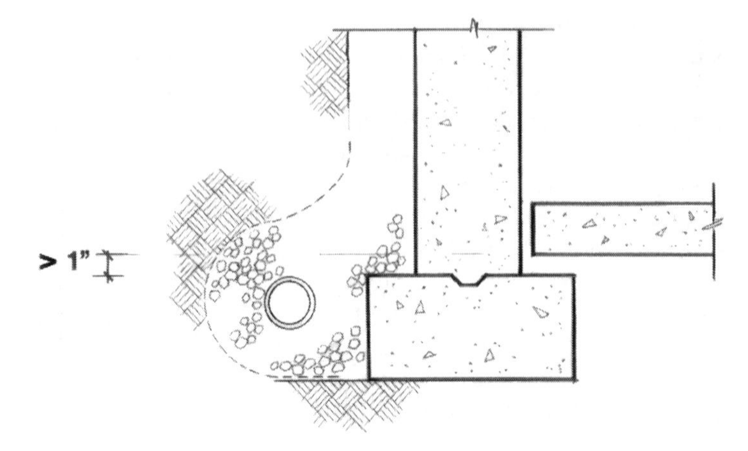

Figure 3.8 When the bottom of the footing is located below the groundwater level, the vapor retarder should go up the inboard face of the foundation to the top of the slab and sealed to it. Where all or parts of the footing will experience exposure to intermittent water immersion, a liquid vapor retarder should be applied to the interface between the footing and foundation. A waterproof epoxy will reduce capillary draw and maintain shear transfer

H. Checklist

1. Check building code requirements before considering dampproofing instead of waterproofing, or elimination of dampproofing.
2. Note that dampproofing is an option only for slabs-on-ground, walls, and planters. Structural slabs are always designed to resist hydrostatic pressure and thus always require waterproofing.
3. If in doubt, choose the most conservative option: waterproofing instead of dampproofing, below-slab rather than top-surface dampproofing.
4. Never apply dampproofing to both sides of a slab-on-ground. It can cause disbonding of the top dampproofing membrane unless the slab is edge vented.

Notes

1 Wayne Tobiasson, "General Considerations for Roofs," *Moisture Control in Buildings*: ASTM Manual 18, 1994.
2 The Building Code requires 6", but 12" provides reasonable safety factor.

4 Basement construction

The term below-grade implies that a habitable space will be constructed below the existing grade or that the existing grade will be increased in height to cover it.[1]

We build basements differently than the above-grade portions of buildings. Basements can be constructed by one of two methods:

1. **Bottom-up construction.** With this method of construction, the earth that will be enclosed by the foundation wall is excavated. Then, the soil retention system is installed and the concrete foundation walls are cast and braced against the soil pressure. The pressure slab is then cast on grade followed by the foundation walls and interior columns. Bottom-up is the most common form of construction.
2. **Top-down construction.** No one would suggest that you start constructing a building by building the roof first and then work down. Yet, for some basements, that is an approved sequence of construction. In this method, the perimeter foundation walls are installed first, followed by the ground floor slab that is cast on the grade. The soil below the slab is then excavated and the subsequent lower basement floors are cast sequentially, finishing with the pressure slab. Columns are inserted in holes drilled into the earth. Generally, as the excavation progresses the foundation walls are braced with cross-lot blocking or rakers and walers as the earth is excavated below the slab exposing the foundation wall. Intermediate upper cellar floor slabs are cast as the earth is excavated to brace the foundation walls and columns. The pressure slab is cast after the excavation reaches the basement level (see Figure 4.1).

A. Excavations and excavation supports

Where space is unlimited, a basement can be formed by simply excavating a hole in the soil, erecting foundation walls, casting the pressure slab-on-grade and backfilling between the soil and the walls (see Figure 4.2).

Where space is limited to a few feet beyond a basement's footprint, the basement can be formed by excavating and installing soil retention systems.

Figure 4.1 Top down construction. In the courtyard of this hospital in New York City where noise and vibration were required to be minimized, the slab at the top was cast on grade. Interior columns were driven to below basement level (some were unintentionally skewed). Earth was then excavated below the slab to the basement level and the mud mat cast on grade. Waterproofing is shown being installed on the mud mat at one of the columns

Figure 4.2 Without property line constraints, this foundation was constructed in an open excavation. Footings and foundation were formed and cast, and positive side self-adhering rubberized asphalt waterproofing is being applied. It is covered with a drainage composite and insulation

A system consists of a network of braces and tiebacks called support of excavation (SOE). *The ASCE Guidelines of Engineering Practice for Braced and Tied-Back Excavations*[2] explains the purpose of SOE. An excavation of soil or rock causes a stress imbalance as lateral support is removed. Deeper and steeper cuts create greater imbalance and greater risk. When the slope of the cut area is not flat enough to retain natural support, an SOE must be constructed to provide support (see Figure 4.3).

If the width of an excavation is limited to the lot lines, the SOE is erected on the lot lines. The waterproofing is pre-applied to the SOE and the foundation wall cast against it. The earth retaining systems can be used solely to retain the earth and act as deep water cut-off walls or they may be designed to also serve as structural foundation walls to support the building above.

The data in Table 4.1 proves that most soil retaining systems are not suitable substitutions for a separate waterproofing system. Some shotcrete system manufacturers claim their systems combine the functions of soil retention, a structural foundation wall, and a waterproofing barrier. These one-size-fits-all systems have not proved to be a complete success despite attempts by (primarily) West Coast builders.

Figure 4.3 Here, property line constraints required this basement to be constructed with a soil-retaining system of soldier piles and lagging with soil nails. Note the dewatering pipes

Table 4.1 Cutoff functions and water-tightness of excavation support of walls

Type of wall	*Potential leakage through wall*	*Potential for piping beneath wall*
Wood, concrete, or steel cantilever sheeting of low height	Deflection opens joints between sheets. Leakage typically 0.1 to 4 gpm per 100 linear feet for low head. Also depending on joint detail.	To avoid piping, penetration below subgrade must exceed $^1/_4$ of exterior head where there is limited depth of pervious layer or more than $^1/_2$ of head for deep pervious layer.
Braced, interlocked steel sheeting	Leakage typically 1 to 10 gpm per 100 lin. ft. of head. Lesser quantity of leakage if movement is minimized, locks are filled and tensile stress acts along the wall in longitudinal direction.	As above, but in potentially "running" soil – non-plastic silt, silty fine sand or narrowly graded sand – piping may occur in a path along the face of the sheeting.
Steel soldier piles with wood or concrete lagging	No impediment to leakage, exterior drawdown and water pressure in active wedge of retained soil can be analyzed by flow net. Inflow depends on soil permeability and recharge.	Not intended for cutoff below subgrade. If exterior head is not greatly reduced by leakage, dewatering to control uplift may be necessary.
Concrete cylinder piles; tangent, secant, or staggered	Equivalent permeability of wall is typically 5×10^{-4} to 10^{-5} cm/sec. Leakage typically $^1/_2$ to 8 gpm per 100' per 10' head, depending principally on quality of joints between cylinders.	Could accomplish partial of complete cutoff below subgrade, but usually not seated into rock. Layout convenient to work around shallow obstructions.
Slurry trench concrete wall	Equivalent permeability of wall is typically 2×10^{-4} to 10^{-6} cm/sec. Leakage typically to 4 gpm per 100' per 10' head. Much influenced by tieback penetration through wall and joint quality. Details of penetrations are important.	Cutoff enhanced by chiseling into underlying rock of by prepositioning grout pipe in wall elements to facilitate grouting in strata beneath wall.

This table illustrates unlikely possibility that the soil retention system can also be used as a watertight barrier for basement construction.

Notes: (1) Leakage through wall is expressed as quantity of inflow per 100-foot length of wall for each 10-foot difference in head across the wall, assuming leakage is not limited by permeability of the retained soil. (2) 1 foot = 0.3 meters; 1 gpm = 6×10^{-5} m^3 /sec.

Source: Published with permission from ASCE.

Soil retaining systems must be restrained to resist soil and water pressure. The soil retaining walls can be tied back to the earth on the outboard side or they must be braced internally where tie backs cannot be installed outside the building footprint. Bracing methods include cross-lot blocking and a system of wales (walers) and rakers. Wales are horizontal members connected to the inner face of the walls. Rakers are inclined struts connected at the top to the walers and braced at the bottom with heel or foot blocks (see Figure 4.4).

You must contend with the inherent problems of internal bracing systems because they penetrate the plane of the waterproofing applied to the soil retaining walls and the membrane under the pressure slab. The following discussion of some of the more common forms of soil retaining systems focuses on the penetrations through the membrane that must be detailed to maintain the watertight integrity of the foundation waterproofing system.

Driven sheet piling consists of profiled steel or wood driven into the ground before excavation. Sheet piling was patented in the 1890s and begun production in the early 1900s. It is manufactured from rolled steel and can be purchased in various shapes. Sheets are driven into the earth and interlocked in a number of ways to form a caisson and limit passage of water and soil particles. Sheets are usually set back from the foundation wall for post-applied waterproofing systems, but may also be covered with plywood to receive a blindside waterproofing membrane system (see Figure 4.5).

Figure 4.4 Rakers with knee braces and supporting sheet piling

Figure 4.5 Steel sheet piling for the deep basement of a 60-story office building adjacent to the Hudson River in New York City

Soldier piles and lagging. This is the most common shoring solution in urban areas. Piles (called "soldier piles") either round or H-shaped structural sections are driven into the earth at least three feet away from the face of the foundation walls. The piles are held in position by being (a) anchored with tie backs into the earth behind them or (b) braced in front of the piles by diagonal beams (called "rakers"). Rakers are anchored to a concrete foot block in the excavation and secured to lateral beams spanning the piles (called "walers"). Wood boards (called "lagging") are fitted between the piles or on one side of them as the earth is removed. This system can be used for post-applied and pre-applied (blindside) waterproofing systems sometimes referred to as one- or two-sided formed construction (see Figure 4.6).

Secant piles are cast-in-place concrete in drilled holes. They are designed as interlocking cylinders with alternate units reinforced by structural steel shapes or cages of reinforcing bars. The inboard faces can be cut back to the faces of the steel pile flanges or a concrete or sand wall installed over them. The reinforced piles are called "primary" piles and the intervening piles "secondary" piles (see Figure 4.7).

Drilled/concrete soldier piles are drilled steel tubular piles that can be filled with concrete. The cylinders are closely spaced and shotcrete or sand walls are installed on their inboard faces.

Figure 4.6 Round soldier piles with lagging are braced with walers and rakers for a below-grade gymnasium

Figure 4.7 Secant piles have been cast and braced with walers and haunches attached to the steel soldiers within the piles. Pile faces will be cut back to the face of soldiers' faces to provide a smooth surface to apply the blindside membrane assembly. Note that haunches must be burned off and penetrations detailed to provide a watertight system. The detail must be designed to protect the membrane from burning

Slurry or diaphragm walls are constructed of bentonite/cement for cut-off walls or cast-in-place concrete for structural walls. They consist of piles bored and filled with a soil/cement mixture (generally bentonite). The soil/cement is then replaced with concrete. These panels have interlocking ends that may incorporate waterstops. They can serve a dual purpose of shoring the site during excavation and acting as a permanent wall when construction is complete. **Precast concrete panels** installed in excavated trenches are a similar form.

Soil nailing with shotcrete. Soil nailing consists of inserting slender reinforcing elements into the soil. The reinforcing is installed into pre-drilled holes and then grouted. Shotcrete (pneumatically applied concrete) is then applied over the surface to act as a rigid facing. Note that the terms soil nails, rock anchors, soil anchors and tie backs are sometimes used interchangeably.

Shotcrete mesh walls are used mostly in western states. Heavy wire mesh is retained in place with soil nails and covered with shotcrete (see Figure 4.8). The skill of the applicator is critical to its success. An inexpensive method sometimes used on the West Coast is to employ the shotcrete not only to retain the earth but to act as a form for the waterproofing. Read "Waterproofing Below-Grade Shotcrete Walls" by Daniel Gibbons and Jason Towle in the March 2009 *Specifier* before specifying this dual-purpose method. The article describes some of the problems with this system.

Figure 4.8 Soil nailing with shotcrete is being installed for the soil retention system of the shallow basement. The machine is shown driving the soil nails

Some of the many other less common types of retention systems include timber shoring, churn drilled soldier piles, wet set soldier piles, and cylinder pile walls.[3]

Where site restrictions do not permit one of the above systems to be used, such as where the vibration of pile drivers is undesirable adjacent to an existing hospital or a high water table is encountered, the primary foundation walls are often constructed by installing caissons of bored secant piles, diaphragm walls, slurry walls, or sheet piling. Alternate concrete piles are reinforced with steel H-shaped piles. This forms a three- to five-foot-thick wall that acts as the foundation wall.

With top-down construction, when the foundation is completed, the first floor slab is usually cast on (1) the grade at the ground level or (2) the wall, which is braced with struts and walers as the earth is excavated, exposing the slurry wall. The slurry walls can then be scarified to produce a reasonably uniform surface.

Cross-lot blocking

Where the site is fairly narrow and neither ties nor rakers are suitable, cross-lot blocking is the preferred system for bracing the soil retaining system. Depending on the depth of the excavation, large pipes are installed to brace the walers at each floor or other practicable vertical spacing. Waterproofing is installed from the bottom up to each waler, and the reinforced foundation cast. Pipes are relocated as required to enable the foundation wall to be constructed successively toward the top. Pipes are removed as intermediate floor slabs replace them (see Figures 4.9 and 4.10).

B. Basement walls

Basement walls (foundations) are designed to resist the lateral pressure of the soil and permanent or intermittent hydrostatic pressure, as well as support the exterior walls of the building above grade. They are generally constructed of reinforced concrete although in shallow basements they may be reinforced masonry.

Where positive side waterproofing is to be applied, the membrane should be covered with protection board or drainage composites and insulation. Drainage composites are not suitable where the water table is above the footing. Where low water tables exist and footing drains are provided, they may be substituted for protection boards. Insulation is required where dictated by code. Where thermal resistance is required, insulation is generally extruded polystyrene, although oriented strand fiberglass can be used in shallow basements. Low density (about 1 pcf) expanded polystyrene can be substituted for protection board. It is adhered to the membrane and becomes sacrificial when backfill settles and internally ruptures the boards.

Figure 4.9 and 4.10 Cross lot blocking is useful for bracing soil retention systems when the opposing basement walls are relatively close together. Here the property lines of parallel streets are one block apart in this below-grade train station in San Francisco. Several tiers of pipe braces are installed to brace the walers attached to the shoring wall. The pipes are progressively removed and reinstalled from the bottom up as the foundation walls are cast

C. Backfilling

The space between the foundation wall and the earth retention system should be backfilled with selected soil, sand, gravel, or low density concrete.

Backfilling is one of the primary causes of membrane failure. Large tree roots, boulders, construction materials, and similar refuse are common in backfill, and often damage membranes. Asphalt felt protection boards (ASTM D6506) generally 1/8–inch-thick offer sufficient protection, although they are not a substitute for careful placement of selected backfill materials.

Sand is better in seismic areas, but generally more expensive than gravel. Backfill materials should be placed in 18- to 24-inch-thick layers and compacted to 85 percent Modified Proctor.[4] Lesser compaction will result in excessive settlement; greater than 88 percent will reduce the drainability. Where the SOE is in close proximity to the foundation walls and structural members project into the normal space to place and compact backfill materials, flowable fill is the preferred material.

Flowable fill, otherwise known as low density cellular concrete fill, is a lightweight concrete ranging from 20 to 60 pcf and 30 to 900 psi, but more often 16 pcf and 50 psi. It is pumped into place and can flow around obstructions. Its use should not substitute for the protection layer. Although it has some thermal resistance qualities, they are of little value when the flowable fill is saturated.

Flowable fill is sometimes used to pad SOE systems and rock-faced excavations to provide a surface on which to apply pre-applied membranes. In this use, it is often termed a sand wall.

Where groundwater is maintained well below the slab, footing drains, and drainage composites or drainage backfills are required to relieve the pressure. The drainage composites and/or insulation generally provide sufficient protection for the post-applied waterproofing or dampproofing. However, where continuous hydrostatic pressure against the foundations exists, the drainage composite is unnecessary and the membrane requires protection only from the backfill.

D. Pressure slabs

The slab-on-grade is designed to resist the uplift pressure of the water when the slab is below the groundwater line. The slab can be increased in thickness to provide sufficient weight to counteract the buoyancy of the water or it can be anchored into position with rock anchors or caissons.

A two- to three-inch mud mat is usually provided on grade to receive the waterproofing membrane. It also serves as a relatively dry working slab. The mud mat is preferable to a tamped gravel substrate, because gravel compaction is never uniform, particularly at rising walls and penetrations. Moreover, penetrations, drains, etc. should be anchored with concrete to ensure that they are held in position during the installation of the pressure slab. Special attention must be paid to the selection of rebar chairs if the tamped gravel alternate is selected.

E. You can't fix what you can't see

Many clients concerned with water infiltration through the pressure slab elect to install a bed of gravel over it for the purpose of installing piping, conduits, etc. They justify this by claiming that it is costly and inconvenient to bury the piping/conduit in thickened sections of the pressure slab or under the slab in trenches. The gravel bed, sometimes in excess of 12 inches thick, is covered with a four-inch concrete wearing surface, in effect sweeping any potential problems under the bed.

Clients put their trust in the waterstops inserted in the cold joints in the pressure slab and the more leak-prone joint between the foundation and the slab. They blind themselves to the fact that the joints are a clear path between the water under the slab and the basement floor, only interrupted by a half-inch-thick bar of rubber or plastic, what in effect is only a concealed caulked joint.

The argument is: If we do get a leak, it will flow through the gravel to a sump and be pumped out of the building. This flawed reasoning ignores the fact that the savings realized by installing the pipes in the gravel bed above the slab is more than offset by the cost of the sump pump(s), the generator, and the sewer tax (presuming that the local utility permits them to dispose of the infiltrating water in its sewage treatment plant).

When a leak occurs, its source(s) cannot be located, which adds to the expense of remediation. Meanwhile, the gravel bed continues to absorb the infiltrating water, the piping and conduits are then in a continuously moist and possibly aggressive environment with a high potential for mold growth. Moisture sensitive floor finishes are threatened and the relative humidity of the space above the wearing slab must be controlled by the mechanical system.

Discuss the hidden costs and risks of this design with clients who wish to employ it.

F. Checklist

1. Consult with the excavation contractor to ascertain the type of SOE system it intends to use. Determine if the components will penetrate the waterproofing.
2. Thicken the pressure slab to absorb piping and shallow pits to simplify waterproofing details.
3. Avoid the system that uses a wearing slab on gravel over pressure slabs to contain pipes and conduits.
4. Provide protection for post-applied membranes.
5. Eliminate drainage composites where flooring drains are absent.

Notes

1 Some Building Codes define a cellar as a habitable space entirely below grade and a basement as a habitable space partly below grade.
2 Committee on Earth Retaining Structures of the Geo-Institute of ASCE, *Guidelines of Engineering Practice for Braced and Tied-Back Excavations* (Geotechnical Special Publication (GSP) No. 74). Reston, VA, 1997
3 Macnab, Alan, *Earth Retention Systems Handbook,* McGraw-Hill, 2002
4 The Proctor compaction test is a laboratory method of experimentally determining the optimal moisture content at which a given soil type will become most dense.

5 Waterproofing

A. When to waterproof or dampproof

Waterproofing can be defined as a material or system capable of preventing water leakage through a subsurface building component under hydrostatic pressure. Dampproofing has a less rigorous function: to resist the flow of water vapor through subsurface building components.

Wherever hydrostatic pressure is exerted against a building component, it is almost always accompanied by water vapor pressure in the same direction: from the wet toward the dry side of the building component. Thus, waterproofing must resist pressure from both liquid water and water vapor flow whereas dampproofing must only resist vapor flow.

You must initially recognize these drastically more rigorous requirements for waterproofing. The decision between waterproofing and dampproofing can be simplified by considering dampproofing as the alternative to waterproofing for subsurface building components. If subsurface components are subjected to hydrostatic pressure, they require waterproofing. If not, they normally require dampproofing. The only exceptions would be made for unoccupied interiors – e.g. an underground parking garage; where relatively high RH, even condensation, could be tolerated. At a minimum, dampproofing is generally required by code. The only exception may be blindside application – e.g. a subbasement cast against rock where waterproofing is not required and dampproofing cannot be installed.

This manual focuses only on waterproofing or dampproofing of subsurface building components that enclose habitable spaces below grade. The three subsurface building components that may require waterproofing are:

1. Structural concrete slabs covered by earth fill or wearing surface.
2. Foundation retaining walls.
3. Concrete slabs-on-ground serving as floors for underground space. (see Figure 5.1)

Framed concrete slabs over inhabited below-grade spaces should always be waterproofed, as they are constantly under the threat of at least temporary

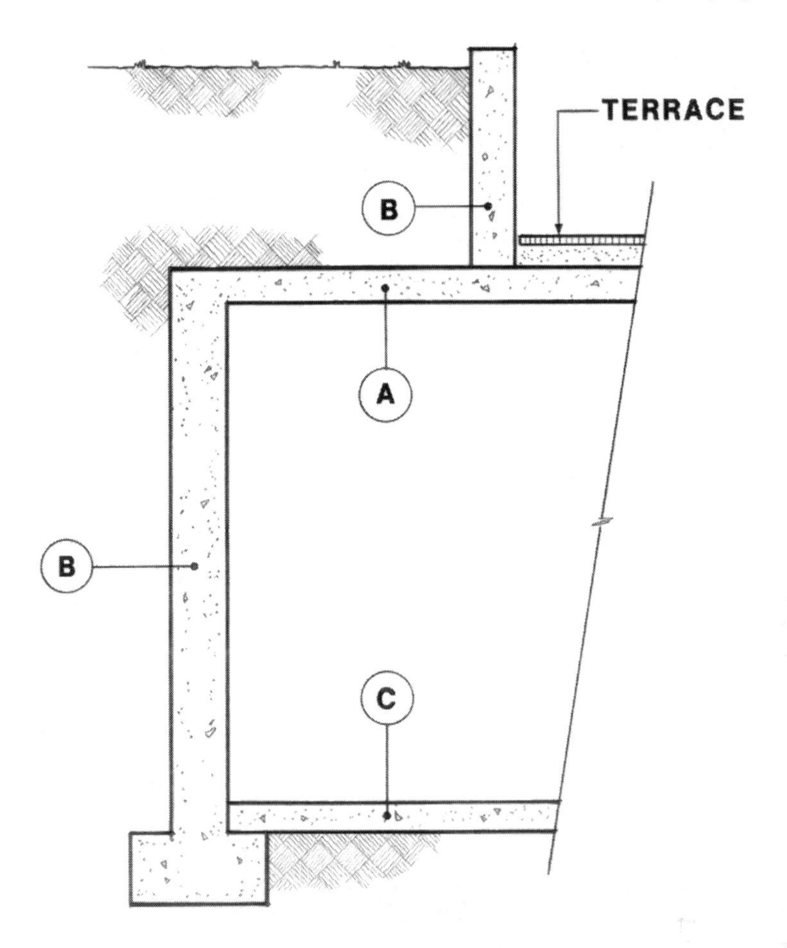

Figure 5.1 Basic waterproofed (or dampproofed) components are (A) framed slabs (always waterproofed), retaining walls (B), and slabs-on-ground (C). Components (B) and (C) may be dampproofed if a slight leak-risk factor is acceptable

hydrostatic pressure. Foundation retaining walls and concrete slabs-on-ground may only require dampproofing (see flow chart, Figure 5.2).

B. Waterproofing types

Waterproofing systems, in descending order of popularity, come in three basic types:

1. Positive-side.
2. Negative-side.
3. Integral.

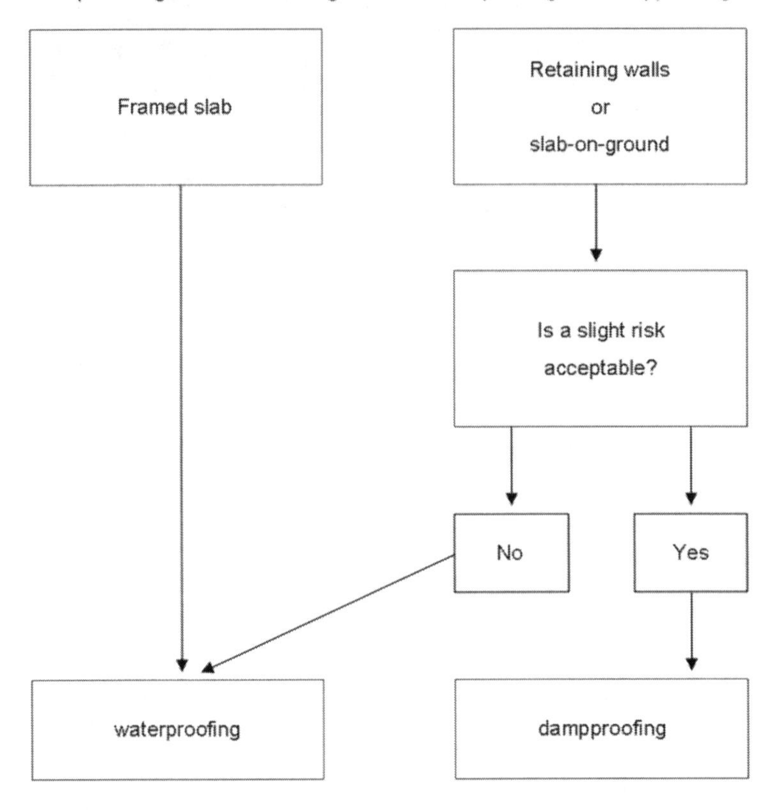

Figure 5.2 The simplified algorithm depicted above decrees a) waterproofing for all framed subsurface slabs or b) retaining walls and slabs on ground through which a risk of leakage is unacceptable. Dampproofing is a residual choice when a slight risk is acceptable

Positive-side waterproofing is applied to the outside (wet) face of subsurface building components. It is the predominant type of waterproofing for new construction, as it prevents a much greater variety of waterproofing problems than negative-side waterproofing does. It comes in a vast variety of subsystems and materials, which gives you myriad options for solving difficult waterproofing problems (see Figure 5.3).

Negative-side waterproofing is applied to the inside (dry) face of underground building components. Easy access to the dry side of waterproofed building components makes negative-side waterproofing the first choice for remedial waterproofing projects. In new construction, its greatest advantage is where blindside application would otherwise be required. As with remedial waterproofing, negative-side waterproofing's major advantage for blindside waterproofing is easy access to the waterproofed surface.

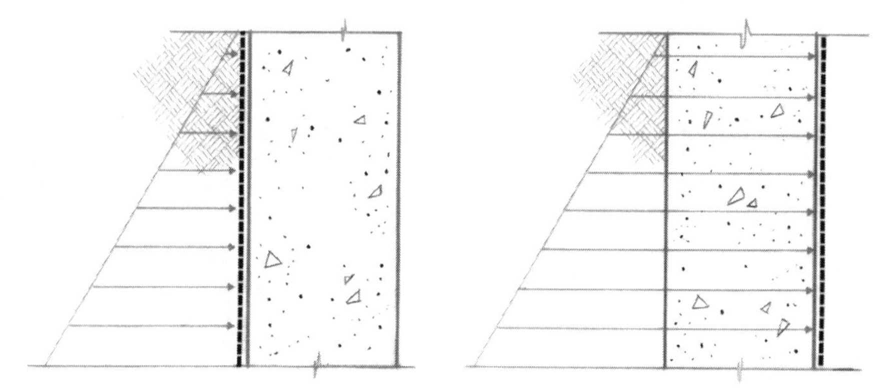

Figure 5.3 Positive-side waterproofing goes on the outside (wet side), free of subsurface building components. Negative-side waterproofing goes on the inside (dry side) surface

Integral waterproofing consists of superplasticizers and other crystal-forming additives enhancing concrete's workability and density. By increasing the hydration of cement, superplasticizers reduce porosity and increase concrete density. So-called chemical conversion or crystalline additives are claimed to form polymers that are hydrophobic and fill the pores as well as inhibit corrosion of rebar. Additives have been used for a century to improve concrete's waterproofing quality. Alum and soap were the earliest, followed by fillers of diatomaceous earth, fly ash, bentonite clay, and pumice.

Despite these physical improvements to concrete, integral waterproofing has nonetheless proven so undependable that it does not qualify as a practicable contemporary waterproofing technique. Admixtures reduce the passage of moisture through concrete, but they are only as good as the concrete itself. Pre-curing and post-curing stresses – from drying shrinkage, temperature change, and structural loading – inevitably crack the concrete and create paths for the passage of water. Since some amount of concrete cracking is inevitable, integral waterproofing is inherently undependable. Furthermore, horizontal and vertical concrete components contain construction (cold) joints that are a common source of leaks. Making these joints watertight is the most critical attribute to ensure a leak-free wall or slab.

C. Designing a waterproofing system (see also Appendix F)

Take the following sequential steps:

1. Review the architectural drawings of the subgrade spaces to evaluate the sensitivity of the occupancy to moisture.
2. Review the governing building code.

3. Review the geotechnical report: groundwater table, water and soil chemical analysis (oil tanks, fertilizers, etc.), expansive clay, rock, and perched water table.
4. Review Support of Excavation drawings where they apply.
5. Review the structural drawings to verify that the structural engineer has designed the below-grade foundations and slabs to resist hydrostatic pressure, rock anchors, and underpinning.
6. Select the appropriate under-slab membrane if hydrostatic pressure exists.
7. Determine if a blindside application is required.
8. Select the appropriate system: pre-applied or post-applied.
9. Select the appropriate membrane.

After reviewing the governing building code, use soil borings and test pits to determine:

- water-table elevation (and the presence of underground streams), tidal and perched water tables;
- deleterious soil chemicals, salt water, petroleum derivatives;
- other hazardous conditions; and
- soil stability.

Determine groundwater level to decide whether to waterproof or dampproof. Depending on when these borings are taken, the water-table elevation used for design will probably require adjustment for seasonal variation. The water-table elevation is likely to be highest in early April in a northern region, where late winter snow melts saturate the ground. Conversely, the water table will usually be lowest in October after summer heat and solar radiation has evaporated much surface water. Soil type is another factor to consider. Sandy, granular soils will normally have higher water-table elevation than less permeable soils containing clay.

Detecting the presence of underground water flows requires supplementary measures in addition to soil borings. Ground with relatively low permeability (e.g. clay) often limits hydrostatic pressure to exceptional circumstances. Conversely, in highly permeable soils (e.g. gravel) you can anticipate more durable worst-case conditions. Corrosive soil chemicals can occur in gaseous, liquid, or solid state. Methane, hydrogen sulfide, and carbon dioxide are among the hazardous gases. Liquids and solids include hydrocarbons, solvents, phenols, and industrial wastes.

Sulfates are a major hazard. They react chemically with Portland cement to form compounds of expanded volume, and this expanded volume can spall concrete and mortar, with progressive cracking and delamination. This destructive process persists while there is a continuing supply of sulfates. The worst situation combines soil sulfates with permeable soils and flowing underground water.

Table 5.1 Classification of subsoil hazards

Classification rank criteria	*Mild*	*Moderate*	*Severe*	*Extreme*
Hydrostatic pressure (Head/Ft)	<3 ft	3–16	16–32	>32
Total acid soluble, SO_4, (%) in soil or fill	<0.24			

D. Systems and materials

Three classes of basement waterproofing membranes are used, depending on when and where they are applied:

1. Plaza.
2. Post-applied.
3. Pre-applied.

They have significant differences between them and very few are suitable for all three types of installations. For example, fewer than a quarter of the manufacturers of plaza waterproofing membranes include resistance to hydrostatic pressure as a physical property. Virtually all of the pre-applied and post-applied membrane systems manufacturers state that they pass the ASTM D5385 requirement for no leaks at a hydrostatic pressure of 207 feet.

Adhesion to the substrate is an important property. Good adhesion reduces the possibility of displacement during backfilling and settlement. It also reduces the migration of water that infiltrates the membrane, localizing leaks and facilitating remediation. Pre-applied membranes for basement walls and under pressure slabs obtain this by either chemically adhering to the concrete or engaging fibers of geotextiles in the wet concrete cast against them. The degree of peel resistance varies from 7 to 28 lbf/inch. Higher resistance is better.

When differentiating between pre-applied membranes, note that they vary markedly in their composition and physical properties. In addition to the method of adhesion, some are single ply, others are built-up. Some single ply sheets are heat welded while others are joined adhesively. This complicates comparison between systems that are claimed to be equal.

One of the most significant factors in selecting a basement waterproofing membrane is its proven durability. In contrast to roofing, waterproofing membranes are inaccessible for remediation or replacement. Roofing is usually guaranteed for 20 years, but waterproofing must last the life of the building, which is 60 to 80 years. Select a membrane that has the greatest chance of performing satisfactorily under hydrostatic pressure, settlement, and attack by aggressive chemicals or seismic events. Pre-applied membranes must also continue to perform when support of excavation systems disintegrate or are displaced. Thus, the most significant yardstick for choosing a system is its history of durability, which should be more than 10 years, and preferably 20 years.

In judging whether a product is sufficiently durable, consider potential liability. Laws or requirements may compel you to use the phrase "or equal" in specifications. In *Untangling the Web of Professional Liability*, Design Professionals Insurance Company (DPIC), an Errors and Omissions insurer, says the phrase means "To possess some performance qualities and characteristics and fulfill the utilitarian function without any decrease in quality, durability, or longevity. No inference that items must be identical in all respects if above conditions are met."

The liability implication is that judges and juries sometimes view a design professional's acquiescence in the use of an item that is not identical to what was specified as a waiver of original requirements.

For a further discussion of this subject see chapter 16.C.1.d., Part 2 Products.

Positive-side materials and systems come in an extremely broad spectrum of materials (described in subsequent chapters). Generically, these systems can be classified as:

- built-up bituminous membranes;
- modified bitumen-sheets;
- prefabricated elastomeric sheets;
- prefabricated thermoplastic sheets;
- bentonite; or
- liquid-applied membranes (LAMs).

Weighing the unique combination of advantages and disadvantages of these materials requires considering many factors, beginning with code limitations and ending with ease of application and feasibility of quality assurance (QA). See Section 5.F. and Chapter 6 for expanded discussion of the design process for positive-side waterproofing.

E. Positive- versus negative-side waterproofing

Positive-side waterproofing is waterproofing on the exterior (wet) side of the building component. It has many advantages over negative-side waterproofing and a much greater selection of available materials. Positive-side waterproofing *includes* the cementitious/crystalline coatings, which is the sole benefit of negative-side waterproofing. Under some circumstances, normally negative-side materials can serve as positive-side waterproofing.

A major advantage of positive-side waterproofing is that it protects against corrosive soils, which can attack masonry, concrete, and even reinforcing bars. Positive-side waterproofing is applicable to all subsurface building components. Because it is located on the inside face of walls and slabs-on-ground, negative-side waterproofing offers no protection against corrosive soil. Even when used as positive-side waterproofing, thus posing a barrier between the soil and the waterproofed component, cementitious/crystalline coatings protect structural concrete slabs less against corrosive soils than membranes do. The unsuitability for structural slabs creates another problem for negative-side waterproofing, as it requires a transition between positive-side and negative-side waterproofing at slab-wall intersections. Combinations of generically different waterproofing systems are normally impracticable (see Section 5.F. Waterproofed Combinations).

Negative-side waterproofing is easier to apply, which suits it for remedial work where positive-side waterproofing failed. It is also easier to maintain than positive-side waterproofing. Easy leak detection is another asset since a leak generally occurs at or near the same location as the defective coating material. Removal of defective material and reapplication of new coating is also easy with negative-side waterproofing. Because it retains moisture within the concrete wall, in effect sandwiching the concrete between a wet (exterior) face and a water-resistant interior face, negative-side waterproofing promotes superior concrete curing. Concrete thrives in the presence of liquid moisture, which promotes continuing strength-producing, solidifying hydration.

Negative-side waterproofing features coatings that seal the interior surface of concrete rather than separate membranes applied to its exterior face. All contemporary negative-side waterproofing systems can be mixed with water, and often with sand; and troweled, brushed, or sprayed as a slurry. Crystalline coatings have largely replaced traditional metallic-oxide cementitious coatings, because of their greater dependability and durability. These solutions of organic and inorganic compounds react chemically with unhydrated Portland cement and free lime, filling capillaries and shrinkage cracks with long chain molecules that crystallize within the concrete. Their superior durability stems from a self-healing property. When cured concrete cracks, the crystalline materials are chemically reactivated and seal the concrete openings and passages.

Offsetting these assets are important liabilities. Negative-side coatings have higher vapor permeability than positive-side membranes do. That limits them to occupancies that can tolerate high interior humidity. Negative-side coatings are also more vulnerable to substrate cracking and fastener damage and do not protect concrete components from corrosive soils.

F. Waterproofed combinations

Only six combinations of building components require waterproofing and dampproofing. These combinations are a logical consequence of basic rules concerning their use. The two basic configurations of waterproofed components can consist of "boxes" – i.e., structural slab on top, foundation wall on sides, slab-on-ground (or earth-supported structural slab) on bottom. Alternatively, they can consist of "Ls" – i.e., foundation and slab-on-ground with no structural slab (see Figure 5.4).

The rules for determining possible combinations are:

- You must waterproof a structural slab. Structural slab is subjected to some hydrostatic pressure from surface runoff.
- If you waterproof the slab-on-ground, you must waterproof the foundation wall. The water table may fluctuate.
- You must dampproof or waterproof the foundation wall *and* the slab-on-ground. Exceptions to this rule are extremely uncommon.

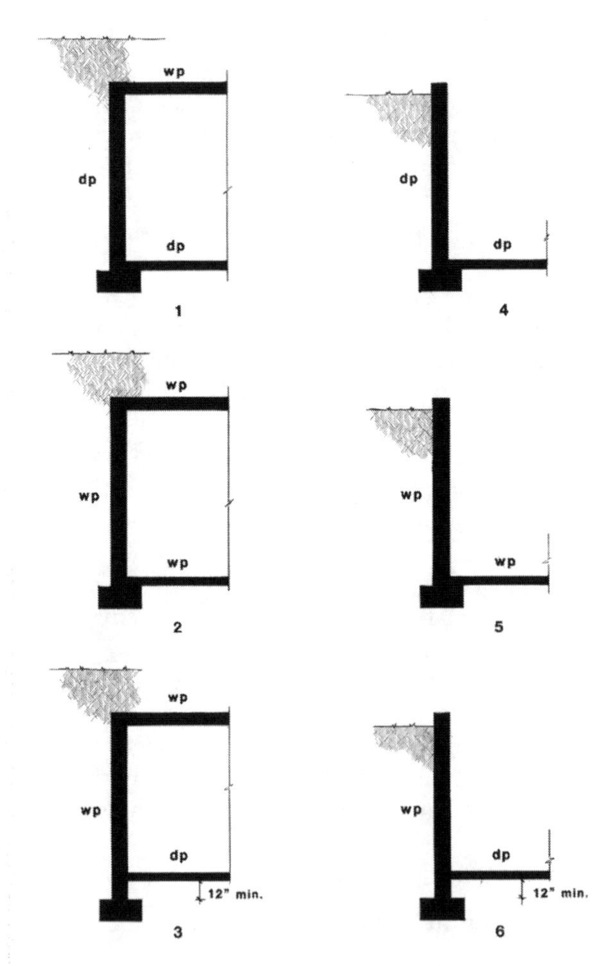

Figure 5.4 The six figures above include all possible combinations of waterproofed (wp) components, except in extremely rare instances where dampproofing (dp) is not required for foundations walls or slabs-on-ground

As depicted in Figure 5.4, permissible combinations are:

1. Waterproofed structural slab plus dampproofed wall and slab-on-ground.
2. Waterproofed structural slab, walls, and slab-on-ground.
3. Waterproofed structural slab and walls plus dampproofed slab-on-ground.
4. Dampproofed foundation wall and slab-on-ground.
5. Waterproofed wall and slab-on-ground.
6. Waterproofed wall plus dampproofed slab-on-ground.

G. Single-system superiority

For the first two combinations, specify the same generic system for all components, with rare exceptions. You may be tempted to combine two or more systems with complementary properties to exploit the best qualities of each material. This is almost always a bad idea. Discontinuities at slab-foundation wall intersections can produce vulnerable joints. Incompatible materials can also create problems. For instance, a plastic sheet membrane for a structural slab and a bituminous membrane for the intersecting foundation wall create a severe joint-sealing challenge. Lack of a positive seal between two incompatible materials can compromise the waterproofing integrity of an entire project.

Combining a waterproofing system with a dampproofing system in the same plane is equally risky (see Figure 5.5). Extending a plaza waterproofing membrane over the edge and down, lapping a dampproofing foundation wall well above the groundwater level is appropriate. Extending a waterproofed membrane up from the footing to some predetermined point below grade, such as two feet above the groundwater level, and then switching to dampproofing is inviting trouble, for two reasons:

1. The possibility of water infiltrating the dampproofing and then migrating downward behind the waterproofing, which may become disbanded.
2. The risk of a fluctuating water table rising above the elevation of the terminated waterproofing.

Among the few exceptions to the general rule against multiple systems is the use of negative-side waterproofing (i.e., cementitious/crystalline coatings) on elevator pits (and other small pits) where waterproofing is installed under a slab-on-ground. Negative-side coatings are also occasionally combined with positive-side waterproofing membranes on structural slabs when the structural slab is above grade.

However, most belt-and-suspenders approaches are highly risky. The most dangerous practice is to apply vapor-impermeable membranes on both

Figure 5.5 Shown above as a foundation with two systems, a membrane system applied to the lower portion of the wall and a spray-applied dampproofing coating applied to the upper portion. The client's faith in the ability of the groundwater to remain within the waterproofed portion was misguided. Leakage occurred shortly after the backfill was installed

sides of a concrete foundation wall. Entrapped moisture will vaporize and delaminate one or both membranes with blisters formed by the expanding vapor trapped at the membrane-wall interfaces.

H. Checklist of considerations

Use a formal, algorithmic process to select a waterproofing system. Avoid the temptation to select a system based on past experience. Prepare a checklist for each building component to be waterproofed. Consider all the conditions affecting each component. Assign a weight to each item. For example, a chemically corrosive soil may outweigh ease of application, thus tilting a decision away from negative-side waterproofing. You can initially eliminate some systems to narrow the range of systems eligible for investigation. For instance, an anti-pollution ordinance may ban hot kettles used for built-up membrane application. Weigh the balance of assets and liabilities, as you would with a financial statement, of the remaining options to make a final selection.

Waterproofing design requires your expertise and careful consideration. You must weigh too many factors for the process to be formulaic. Consider these factors:

1. Code requirements.
2. Occupancy.
3. Hydrostatic pressure.
4. Water table.
5. Soil characteristics.
6. Substrate stability.
7. Construction sequence.
8. Durability of material(s).
9. Ease of application.
10. Risk versus cost.

1. **Code requirements** set mandatory parameters that narrow your options. As a notable example, the ICC Building Code enumerates acceptable waterproofing systems that are approved for use on slabs and walls and concludes with "or other approved materials." That vague phrase leaves the door open to seek code officials' approval for a material outside of the stated scope. All codes contain provisions permitting the exercise of code officials' discretionary approval, which you must seek if your system selection is not explicitly approved by the code.

Volatile organic content (VOC) regulations established by many states thus far are a related consideration. These regulations can eliminate the choice of some waterproofing materials, especially solvent-based adhesives, and primers. Pollution regulations thus require investigation early in the design process.

2. **Occupancy** is a vitally important design factor. Leak-risk tolerance and humidity sensitivity are two system-eliminating factors that can rapidly reduce your choices. Occupancies with no risk-tolerance include libraries, museums, rooms with a computer or other electronic or electrical equipment, and medical facilities. These same occupancies often require tight humidity control as well. Occupancies with audio-visual equipment, or art storage or exhibition, virtually eliminate negative-side waterproofing, and usually bentonite, from consideration. Humidity-sensitive occupancies require positive-side membranes with low vapor permeance.

3. **Hydrostatic pressure** is, by definition, the basic factor in the choice of a waterproofing system. Aspects demanding consideration are intensity, duration, and nature of hydrostatic pressure – i.e., whether the pressure is continuous or intermittent and whether the water is stationary or flowing. A fast-flowing underground stream eliminates bentonite. Like asphalt, bentonite deteriorates from the erosive action of flowing water, but much more rapidly.

The nature of the hydrostatic pressure is also an important factor. Under intense, continuous hydrostatic pressure, membranes with low moisture absorptance perform well. High moisture absorptance causes membrane swelling, disbonding, and consequent wrinkling, as the expanding sheet buckles under compressive stress. Without substrate support, a wrinkled sheet membrane is subjected to a vastly increased risk of puncture.

Intense hydrostatic pressure poses other hazards. It can force membranes into voids in the concrete, where cracking under flexural stress can ultimately rupture the membrane and admit leaked water. Because its material depends upon a constant supply of water, making it a good material where hydrostatic pressure is constant, bentonite is a poor choice where hydrostatic pressure is intermittent (see Chapter 12).

On negative-side waterproofing, intense hydrostatic pressure forces water into voids formed by tie-rod holes, cold joints, and rock pockets, from which it exerts pressure on the negative-side coatings. This intensified pressure turns minor imperfections into more probable sources of leakage. However, these defects can easily be repaired from the inside.

Where hydrostatic pressure is intermittent and fast-flowing underground streams can erode bentonite clay systems, negative-side waterproofing may prove the best choice, but only if the occupied space is insensitive to high humidity – e.g., a parking garage or mechanical room. For expanded discussion of hydrostatic pressure, see Chapter 2.A.

4. Water-table level must be determined by accurate soil borings with results adjusted for seasonal variation. In the Northern hemisphere, the water table is normally highest around the vernal equinox (after spring thaws have saturated the ground). Conversely, the water table is normally lowest around the autumnal equinox (following the summer evaporation of surface moisture). Thus, soil borings taken in July, August, or December may not accurately depict the most severe water-table elevation.

5. Soil characteristics include both chemical and physical properties that affect waterproofing. For negative-side systems, which leave foundation walls unprotected, pH values must be determined to assess the acidity or alkalinity of the soil. Acids or alkalis in groundwater can accelerate deterioration of concrete and steel reinforcing bars. Other corrosive soil chemicals include salt water, salts, sulfates, calcium hydroxides, oils, and tars. Salt water corrodes reinforcing bars in concrete.

Sulfates are especially hazardous in concrete. They react with Portland cement to form chemical compounds with expanded volume. The resulting internal compressive and shearing stresses can spall the concrete, often exposing steel reinforcement to corrosive attack from acids and other soil contaminants. The most corrosive of these sulfates are magnesium, sodium, and ammonium. Though ammonium sulfate does not occur naturally in soil, it may be present in farm land, where it is used as a fertilizer.

Physical soil conditions are also a concern. Clay soils with low permeability tend to limit underground hydrostatic pressure, whereas highly permeable granular or sandy soils, such as Florida's, maximize hydrostatic pressure.

6. Substrate stability refers to the propensity for cracking in underlying masonry or other waterproofed components with many construction joints. Expansive soils encountered in the West and Southwest, and peaty soils, can produce rising or settling footings, which induce stresses that crack the footings and foundation walls. Over substrates vulnerable to cracking from any source, the membrane must be capable of elastically elongating or resealing. Elastomeric and modified-bitumen sheets with high breaking strains work best over unstable substrates.

7. Construction sequence requires close attention to prevent vulnerable materials from being exposed to the elements when long delays occur during construction. Bentonite clay panels must be shielded from rain and construction water. During cold-weather construction schedules, waterproofing systems must be capable of resisting subfreezing temperatures when exposed for more than a week. Membranes with low resistance to ultraviolet radiation on brick shelves can deteriorate if exposed to sunlight for intervals as short as a month before they are shielded by the bricks.

Construction sequence can sometimes put waterproofing under anticipated stress. A deep foundation wall for a two-level basement may rely on the lateral support of an intermediate structural slab, shear walls, or shafts. If the foundation wall is waterproofed before these structural elements are cast, the foundation wall may deflect excessively, with disastrous consequences for the membrane. See Chapter 17.A. for more discussion about sequencing problems.

8. The durability, manufacturer's reputation, and limited warranty of each system under consideration are important factors. To assess a system's durability, ask these questions: is the product manufactured by the seller or is the seller merely a distributor? (Dun & Bradstreet can provide this information if a direct query fails.)

- What is the manufacturer's reputation? Were previous product introductions or changes successful? Is the company supportive or combative about honoring its warranty when a product's performance declines?
- Does the product have a history of successful performance for at least 15–20 years under comparable conditions?
- Has the waterproofing component maintained a consistent or "improved" formulation for at least 10 years?

9. Ease of application is a relatively minor factor. Facilitating the application may result in better workmanship. Application of modified-bitumen membranes is, for example, easier than dependable application of LAMs. On a project where other factors are roughly in balance, ease of installation could tip the decision toward modified-bitumen rather than a LAM.

10. Risk versus cost, the final design factor, is probably the easiest to evaluate. You should always minimize risk, regardless of cost. If a client must cut construction costs, the waterproofing system is among the last places to economize. You cannot design a system with no possibility of failure, but you should err on the conservative side when evaluating design factors, matching them against a manufacturer's claims before committing to a less expensive system.

As an example of prudent design, the waterproofing designer of an underground space housing computers or rare books should recognize the possibility of flash floods and water-main or sewer-line breaks. Those types of incidents can load the waterproofing system more severely than normally anticipated hydrostatic pressure from groundwater or perched water tables would. Water-main breaks are quickly reported and repaired. Sewer-line breaks, on the other hand, may either be tardily repaired or never repaired. Water-line breaks signal an immediate problem, but sewer-line breaks and their consequent leakage may not be obvious, beyond lightening the load on the treatment plant.

I. Positive-side waterproofing checklist

1. Check the soil analysis report for chemicals and groundwater level. Require the waterproofing manufacturer's representative to certify that he has read the soils report.
2. Verify the applicator's credentials.
3. Determine if the groundwater level will be sufficiently maintained below the slab on grade to eliminate under-slab waterproofing.
4. Compartmentalize loose-laid sheets, if they are used.
5. Specify substrate preparation in the concrete division.
6. Select positive-side waterproofing for corrosive soils and salt-water environments.
7. Limit water absorption of membrane materials to two percent if practical, and in no case higher than five percent (per ASTM Test Method D95).

J. Blindside waterproofing checklist

Select a system that will bond to the concrete foundation for blindside waterproofing where vapor control is critical.

K. Negative-side waterproofing checklist

1. Ban installation of negative-side waterproofing in new construction under *any* of the following conditions:
   a. presence of corrosive soil chemicals
   b. exposure to freeze-thaw cycling
   c. a low interior-humidity requirement

2. Require manufacturer's approval of applicator.
3. Require pre-installation inspection of substrate surfaces.
4. Cite surface preparation requirements.

6 Positive-side waterproofing systems

Positive-side waterproofing (called post-applied) is applied to the outside (wet) face of subsurface building components, in contrast with blindside (post-applied) and negative-side waterproofing applied to the inside (dry) face of subsurface walls and slabs (see Figure 5.3). For new construction, positive-side waterproofing is much more widely specified than negative-side waterproofing, which is the most popular system for remedial work.

Additional uses for positive-side waterproofing include the inside faces of water-containment structures – e.g., planters, swimming pools, tanks, fountains, and dams.

Waterproofing materials were formerly limited to built-up bituminous membranes and negative-side cementitious coatings. LAMs, self-adhering sheets, and bentonite, all introduced in the past half century, have become more popular than the older materials. Negative-side waterproofing is the primary system for vehicular and utility tunnels where the outside is inaccessible, and also for liquid storage and swimming pool structures.

Positive-side waterproofing prevents a much greater variety of waterproofing problems, with fewer limitations, than negative-side waterproofing does. It also comes in an overwhelming variety of subsystems and materials, which gives one many tools for solving special waterproofing problems.

In addition to the cementitious coatings used for negative-side waterproofing, positive-side waterproofing materials include bentonite, generic categories of single-ply sheets (elastomeric, thermoplastics, modified bitumen); built-up membranes comprising felts and fabrics, and LAMs. See Chapters 9–12 for a detailed discussion of each membrane type.

Positive-side waterproofing offers several notable advantages. Unlike negative-side waterproofing, it is applicable to all subsurface components. As noted in Chapter 5, negative-side waterproofing is impractical for framed structural slabs. This disqualification spoils the continuity of negative-side waterproofing as a total system. It always requires a transition between positive-side and negative-side waterproofing at joints between walls and structural slabs. As explained in Chapter 5.F and Chapter 6.B–F, combinations of generically different waterproofing systems or materials are generally inadvisable.

As a second advantage, positive-side waterproofing protects against corrosive soils, which can attack masonry, concrete, and even steel reinforcing bars. Corrosive soils can thus pose a threat to the system's structural integrity. With negative-side waterproofing, you implicitly accept these risks. (See Figure 6.1 for an algorithmic flow chart for preliminary system selection.)

A. Positive-side design factors

Positive-side waterproofing comes in a variety of materials:

- bentonite clay
- modified bitumen sheets
- LAMs
- built-up bituminous membranes
- prefabricated elastomeric sheets
- prefabricated thermoplastic sheets
- cementitious or crystalline coatings

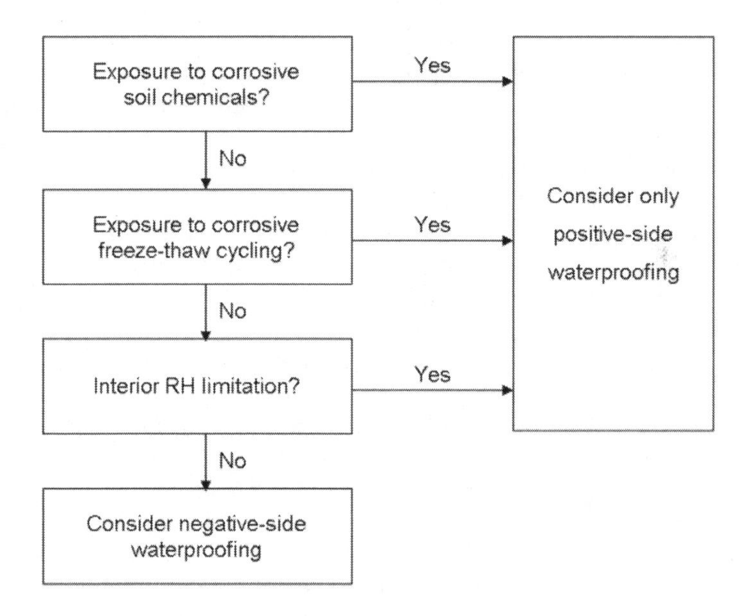

Figure 6.1 Algorithm diagram shows preliminary design elimination process for choosing between negative-side and positive-side waterproofing systems for new construction. For remedial work, where positive-side waterproofing is impracticable, negative-side waterproofing has a stronger case than is indicated by the algorithmic procedure depicted above

These materials each have unique combinations of assets and liabilities. Their selection requires delicate balancing of numerous design factors, notably the following:

- code limitations
- your past experience with a particular system
- geometric complexity
- feasibility of quality assurance (QA)
- resistance to soil chemicals
- ease of application
- resistance to design hydrostatic pressure
- vapor permeance
- substrate stability
- accessibility for installation

Code limitations, for both pollutants and safety, are the first obstacle considered in an efficient design process because they eliminate some options. Hot-applied built-up systems, traditionally the most popular, are increasingly banned by pollution regulations in major metropolitan areas plagued by air pollution. The most effective built-up waterproofing membrane, coal tar pitch, is, unfortunately the most heavily polluting, which has resulted in its discontinued use.

Eliminating a hot-applied built-up system may leave you with the alternative of a cold-applied built-up membrane. Yet, those may also be banned. Solvent-based asphalt cutbacks, used as the primer or adhesive in many cold-applied built-up membranes, may violate VOC regulations in more rigorously regulated jurisdictions.

Your past experience with a system ranks high as a design factor, but with caveats. You should not consider a system until it has provided at least 10 years of service unless your client expresses written willingness to accept the additional risk associated with a newer system. On projects designed to resist high hydrostatic pressure, no material should even be considered unless it satisfies the 10-year minimum service-life criterion. When specifying waterproofing material, even material that has performed satisfactorily, make sure that the formulation has remained essentially unchanged. If it has not, consider the reformulated material as an essentially new material, or system.

Geometric complexity can pose problems for some waterproofing systems; notably, the felt-laying, hot-mopping operations of built-up membrane application. With their more readily adaptable application techniques – rolling, brushing, squeegeeing, or troweling – LAMs are readily installed on irregular substrates. Judged by this adaptability criterion, prefabricated, single-ply sheet membranes fall somewhere between built-up and liquid-applied membranes.

Feasibility of quality assurance is closely correlated with the previously discussed factor. QA denotes measures undertaken by the *client*, as opposed to the *contractor*, to assure correct system installation. The feasibility of dependable QA requires an established set of well defined application procedures that are empirically proven to produce waterproof components. A new, experimental waterproofing system is unlikely to have proven, well defined procedures. Many waterproofing projects are not supported by rigorous QA programs. Some waterproofing failures result from such inexcusable lapses as the omission of protection boards that shield waterproofing membranes from damage during backfill operations and subsequent construction above them. These projects appear to have been unsupported by QA.

You should also investigate what quality control (QC) measures will be taken by the contractor to assure correct system installation. Two parties looking for flaws in construction practice are better than one. QA is, however, much more important than QC, as the client has a greater stake in good construction practice than the contractor does.

Waterproofing installers should be certified or approved by the manufacturer. Shop drawings and a pre-application conference are mandatory. The first day's work should be inspected by a technically qualified manufacturer's representative, rather than a salesperson. Depending on project complexity, periodic and final inspections should be made. Each lift of wall waterproofing should be inspected before it is backfilled. The manufacturer's representative should also witness flood-testing and repairs. See Chapter 16 "Specifications."

Resistance to soil chemicals can focus your attention more intensely on feasible solutions early in the design process by eliminating some systems and materials. Laboratory analysis of site soil samples can identify hazardous chemicals. The presence of corrosive chemicals may eliminate one of the most popular positive-side waterproofing materials – bentonite clay – which breaks down in the presence of some soil contaminants. Consult with the manufacturer of a product or system under consideration about its vulnerabilities.

Chemical contaminants that may be present in varying concentrations include:

- soil poisoners, (usually petroleum based)
- fertilizers
- toxic wastes
- salt water (including de-icing salts)
- waste oil

Ease of application may have decisive local implications. In a large metropolitan area, the prospects of finding competent local contractors for any system is greater than in other areas where specialized expertise for applying some waterproofing systems such as LAMs, may not exist.

The remaining design factors – **hydrostatic pressure, vapor permeance, substrate stability**, and **accessibility for installation** – require little discussion.

Low vapor permeance may require some research for suitable materials if the waterproofed space requires controlled humidity – e.g., for rooms housing computers or other electronic equipment and even carpeting, which can be damaged by high humidity. As one notorious example, manufacturers' literature for bentonite products generally contains scant guidance on their vapor permeance. You may have to request that information.

Accessibility for installation can be a determining factor if there is inadequate space to install membranes. Generally 30 inches is adequate for most post-applied systems.

B. Single versus dual systems

As a general rule, you should specify a single generic waterproofing system. Some conditions may favor a combination of systems as the best design solution. Certain membrane materials are more suitable for horizontal than for vertical substrates, especially LAMs, built-up bituminous membranes; rubber; and PVC sheets. In those situations, you may choose a combination of two different membrane materials to take advantage of their complementary properties. However, you must exercise care in not attempting to join incompatible materials.

As an example of the advantages of combinations, on projects with positive-side membrane waterproofing elevator pits and other small pits can be waterproofed with negative-side, cementitious or crystalline coatings. Unlike most combinations, which require an awkward transition joint between two dissimilar materials or have an incompatibility between these materials, this system features a practicable overlap of the two different waterproofing systems (see Figure 6.2). Usually, the difficulty of detailing a dependable transition joint between two generically different materials – especially at wall-slab junctures – weighs against a dual system. However, a common system of applying hot rubberized asphalt membrane on a plaza slab and lapping it over a self-adhering rubberized asphalt membrane on a foundation wall is widely used by torching off the polyethylene film on the foundation membrane to achieve a bond.

Combining a waterproofing system with a dampproofing system in the same wall plane can be risky unless the materials are generically compatible, a plaza slab membrane can be extended over the edge of a foundation and down the wall, lapping a dampproofed foundation wall well above the groundwater level. But it is inviting trouble to extend a waterproofing membrane up from a footing to a predetermined height above the groundwater table for a transition to dampproofing.

The most dangerous combination, however, is the belt-and-suspenders' approach of applying vapor-impermeable membrane systems to both sides of a concrete wall or slab. When the entrapped moisture evaporates, it can create blisters, delaminating one or both membranes from the wall. Bentonite panels against a membrane-waterproofed wall are an equally

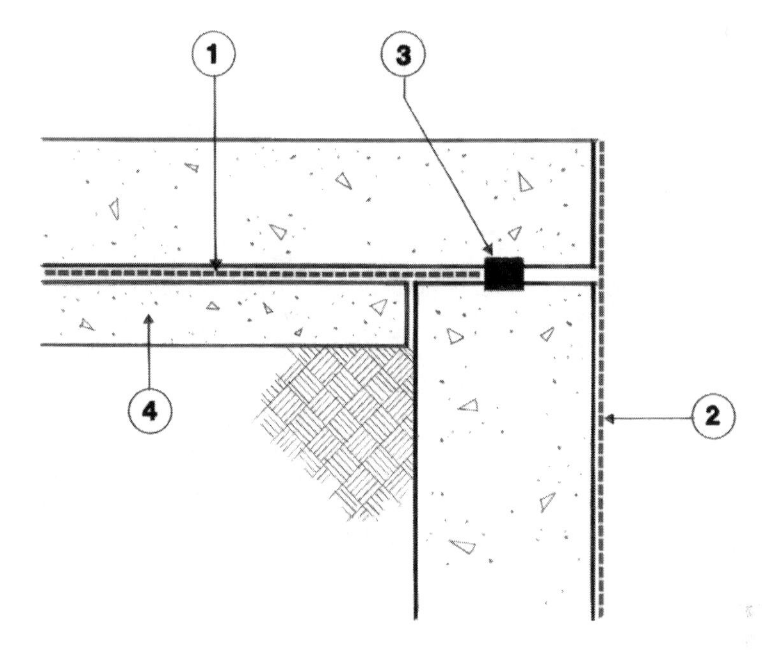

Figure 6.2 Elevator pits and other small pits offer a rare opportunity for successful combination of positive-side and negative-side waterproofing. An alternate method of waterproofing pits uses positive-side waterproofing for both slab-on-ground and pit. Notations: (1) waterproofing membrane, (2) crystalline slurry coat, (3) water stop, (4) mud slab

disastrous combination. The nails anchoring the bentonite panels will eventually corrode and admit leak water through the nail holes in the perforated membrane.

C. Adhered versus unadhered

Your most important decision is choosing between two basic concepts: fully adhered or unadhered (i.e., loose-laid or spot-bonded) membranes. Many prefabricated membrane materials – PVC and other thermoplastic materials – can be loose-laid and compartmented, spot-bonded (partially adhered), or fully adhered to the substrate. All other membranes must be fully adhered.

Compared with fully adhered waterproofing systems, loose-laid or spot-bonded systems offer these advantages:

- less dependence on substrate preparation
- reduced or eliminated membrane-splitting risk

- superior joint seams on thermoplastic sheets
- less sensitivity to climatic conditions

Less dependence on substrate preparation is beneficial because adhered systems require a smoother, more uniform finish, and a much cleaner and drier surface than non-adhered systems do to assure high bond strength at the membrane-substrate interface. Membrane adhesion to a deck can be impaired by dust, grease, liquid moisture, and some curing and form release agents. Liquid moisture on a concrete deck boils when hit with hot bitumen, and the resulting voids in the mopping layer can permit lateral movement of water. Some curing agents, particularly sodium silicate, can also spoil adhesion. They form a crystalline molecular structure that fills concrete pores, and glazes surfaces. This glaze is often visually undetectable. Unlike some acrylic and alkyd resin-based curing agents, which flake off in 14 to 28 days, sodium silicate requires sand blasting to provide an acceptable surface. Pinholing and complete loss of adhesion of liquid-applied membranes can often be traced to the presence of sodium silicate curing agents.

Adhered membranes require priming of concrete (or parged masonry) surfaces to provide tack and absorb dust, which can also impair adhesion (see Figure 6.3). VOC regulations have forced many waterproofing manufacturers to abandon solvent-based coatings and primers in favor of water-based coatings and primers. Water-based products do not "wet" or penetrate the surface as well as solvent-based materials. Since water-based products are not as "forgiving" as solvent-based materials, they must be applied to cleaner, smoother surfaces, which require more rigorous preparation. One exception is water-based epoxy primers used where substrate moisture is unacceptably high.

Specifications for preparing and repairing concrete to be waterproofed can be found in ASTM D5295 and ACI 515. Tie-rod holes, static and dynamic cracks, honeycombing, rock pockets, ridges created by formwork kickouts, and similar surface irregularities require correction. Irregularities are remedied by grinding or chipping. Avoid sanding, because it polishes the concrete and closes the pores to primers and coatings. Select form release oils that do not inhibit bonding or proper penetration of waterproofing constituents.

Cracks must be filled or detailed by stripping as required by the membrane manufacturer. According to some authorities, 60 percent of drying-shrinkage cracks will occur in the first 28 days.

On masonry foundations, the Brick Institute of America and National Concrete Manufacturers Association recommend that waterproofing and dampproofing be applied to a half-inch thick cement mortar parge coat rather than directly to the masonry. These parge coats are required because mortar joints are not always full or struck flush, and broken corners of CMU are not always patched. A parge coat provides a smooth surface for

Figure 6.3 Adhered membranes (built-up, modified-bitumen, etc.) require priming of concrete substrates to absorb dust and provide a tacky surface assuring tight adhesive bond at the membrane-concrete interface. (TC Mira DRI)

the waterproofing or dampproofing. It also adds some water resistance to the foundation.

Some waterproofing systems require cementitious cants, coves, or fillets at reentrant angles. Some require bevels at external edges. Include these in the concrete trade construction documents or they may incur an extra charge.

Unadhered membranes present **reduced splitting risk** because substrate cracking is propagated directly up into adhered membranes. If, for example, a 1/8-inch crack occurs in a concrete substrate topped with an adhered membrane, the strain (elongation divided by length) may be 100 percent or more. A loose membrane can bridge over the crack, experiencing a strain of one percent or less, well within the breaking strain of any common waterproofing membrane. (Vulcanized rubber membranes generally have breaking strains of 300 percent or more. Built-up membranes, however, have breaking strains around two percent.)

Because of their vulnerability to splitting, adhered waterproofing membranes should be selected for their ability to accommodate substrate cracking. Membranes can accommodate substrate cracking through two mechanisms:

1. "bridging" the crack without rupturing, or
2. self-healing, if ruptured.

Bentonite clay offers a unique, self-healing solution to this problem; see Chapter 12.

The superiority of joint seams sometimes claimed for unadhered systems, is limited to thermoplastic sheets, which are predominantly PVC, TPO, and KEE. The heat-welding of seams in a thermoplastic sheet produces a joint as strong as the sheet material itself. This advantage for unadhered systems does not apply to EPDM, the only other material suitable for loose-laid membranes. The adhesive-based seams used for EPDM membranes generally lack the strength of the sheet material and require a cover tape for seam reinforcing.

This argument for the superiority of unadhered systems is further weakened by adhered system alternatives, which include polymer-modified bitumen as well as PVC and EPDM. As one such alternative, a two-ply system with self-adhered modified-bitumen sheets, laid one-on-one with staggered joints, virtually nullifies the claimed superiority of thermoplastic loose-laid systems. A two-ply adhered system thus constructed offers redundancy. In contrast, loose-laid, single-ply membranes require greater care during installation to assure the integrity of their seams, which provide the first and only line of defense.

Less sensitivity to climatic conditions is an irrefutable advantage for unadhered membranes. Moreover, large plastic and rubber sheets are readily installed on horizontal surfaces but are less practical for vertical surfaces.

Compared with unadhered membranes, adhered membranes offer two big advantages:

1. Easier leak detection.
2. Lower probability of membrane slippage during backfilling operations or subsequent settlement of poorly compacted fill.

Easier leak detection is the upside of adhered membranes' greater splitting risk. When water leaks through a crack in a slab or wall with an adhered membrane, you can reasonably assume that the membrane rupture closely coincides with the crack's locations.

In a loose-laid membrane, however, the appearance of the leak in a structural slab soffit or slab-on-ground may tell you little about the location of the water-admitting membrane defect. It may be located 30 feet or more from the interior leak location. Knowing the location of the membrane defect normally facilitates remedial work. The detective work in locating a leakage from unknown source(s) in a membrane area of 1,000 square feet can take more time than the repair itself.

Easier leak detection is accomplished by compartmentalization, at least on horizontal surfaces. The substrate is divided into squares by applying adhesive bands in a 10 foot grid. With the membrane adhered to the bands, leaks are confined to the gridded compartments. Where a LAM is called for by other design factors, a system suggested by Charles Parise, which combines a LAM with a loose-laid rubber sheet, exploits the unique advantages of each system.

On one hand, the compartmentalized adhered membrane facilitates detection of the leak's location by confining it to an area much smaller than the area that must be investigated when a loose-laid membrane leaks. On the other hand, the loose-laid membrane reduces the probability of a leak's occurrence. A loose-laid system reduces the number of seams, which are the most vulnerable spots in a waterproofing membrane. It also exploits the superior dependability of welded seams, at least when thermoplastic sheets are specified. For a detailed discussion of seaming techniques, see Chapter 10.

An adhered system's **greater resistance to slippage** is attributable to its greater frictional resistance to gravitational force parallel to the wall surface. Perpetual gravitational force stretching a partially adhered membrane can exert gradually increasing concentrated stresses that can ultimately result in adhesion failure at spot-bonded locations. Fully adhered membranes are subjected to lighter, uniform shearing stresses in the adhesive layer. Built-up membranes are usually installed in six- to eight-foot strips laid vertically and back-nailed to wood nailers to prevent slippage or sagging.

An even bigger problem for non-adhered wall membranes is downward frictional force exerted by backfill settling adjacent to the wall and transmitted through the protection boards or drainage composites sandwiched between backfill and membrane. This frictional force results from the shrinkage of the backfill materials and the natural consolidation of soils as they progress from their less-dense condition as backfill to a permanent, "undisturbed" condition. (This soil consolidation itself is a consequence of gravitational force.)

Another controversy between proponents of adhered and proponents of unadhered membranes concerns membrane adhesion to lot-line walls and mud slabs. According to proponents of unadhered single-ply membranes, adhered membranes applied to lagging, lot-line walls or mud slabs may fail to bond to concrete cast against them. As a consequence, they allegedly share the same untraceable leak problem of loose-laid membranes. See Chapter 8 "Blindside waterproofing."

Proponents of adhered systems counter that when concrete is cast against some types of bituminous membranes, the heat of hydration liberated by hardening concrete softens the bitumen enough to create a bond. Membranes benefitting from this phenomenon are built-up modified-bitumen sheets without polyethylene facer films.

Other membranes used for pre-applied application include HDPE sheets coated with adhesive and sheets faced with geotextiles. These are discussed at greater length in Chapter 8. One method of exploiting this phenomenon, and thus reducing the probability of lateral leak water flow between built-up membranes on mud slabs and slabs-on-ground above, is proposed by Dorothy Lawrence, who claims consistent success from its use. It comprises these steps:

1. Install the first ply on the dry, unprimed mud slab. Use adhesives in spot locations to hold the ply in place during subsequent installation operations.
2. Install the remaining plies in the normal manner (i.e., hot-mopped or cold-adhered, depending on the specified system).
3. Cover completed membrane with protection boards bonded to the membrane. (Orient protection boards with polyethylene slip sheets on top and lap edges. Omit protection boards on vertical surfaces.)
4. Immediately before casting of slab-on-ground, remove polyethylene film and then prime protection board surface.
5. Place welded wire fabric or other slab reinforcement.
6. Cast concrete.

Because the first ply is isolated from continuous contact with the mud slab, this method minimizes chances of mud-slab cracks propagating upward into the membrane. Subsequent saturation of the mud slab will disbond the membrane, but the hardening concrete's heat of hydration should raise the bitumen temperature enough to adhere the protection board-membrane laminate to the slab above.

Continuity in the adhesive layer between membrane and substrate is essential in an adhered system to prevent entrapment of water between the two surfaces, where it can work destructive mischief. Entrapped water can be absorbed by both membrane and substrate material with consequent swelling of the materials and expansion of disbonded areas. This degenerative process can ultimately produce blisters in some membranes, and these blisters can destroy the system's waterproofing integrity.

To prevent this degenerative process, you should specify membrane materials with low water absorption. David Adler and Stephen Ruggiero of Simpson Gumpertz & Heger recommend a maximum limit of two percent. Systems with two-to five-percent water absorption are generally satisfactory. They suggest that materials with absorption greater than five percent are unsuitable for waterproofing.

D. Summary

You should initiate the material selection process with an algorithmic process to shorten the list of options from among the tremendous variety of waterproofing materials. First, eliminate materials that are unavailable or banned by regulation. Environmental regulations, for example, may exclude both hot-and cold-applied membranes, and some LAMS. (See Table 6.1 for other design factors that eliminate, weigh against, or favor certain systems.)

Subsequent chapters detail the specific advantages and disadvantages of various positive-side waterproofing materials cited in this introductory chapter.

E. Checklist

1. Check the soil-analysis report for chemicals and groundwater level. Require waterproofing manufacturer to certify that a representative has read the soils report.
2. Select positive-side waterproofing for corrosive soils.
3. Determine if the groundwater level is sufficiently below the slab-on-grade to eliminate under-slab waterproofing.
4. Verify the applicator's credentials.
5. If loose-laid sheets are used, compartmentalize them.
6. For blindside waterproofing where vapor control is critical, select a system that will bond to the concrete foundation.

Table 6.1 Preliminary system selection process

Design factor	*Effect on system selection*
Rapidly flowing water	Eliminate bentonite
Corrosive soils	Eliminate negative-side waterproofing, bentonite
Blindside waterproofing	Eliminate LAM, PVC
Subfreezing temperatures	Eliminate negative-side waterproofing and cold-applied emulsions; extreme temperatures make built-up and self-adhering systems doubtful
Tight construction schedule	May eliminate LAM, which requires very dry substrate
Constructed work site, small building elements, VOC compliance	Eliminates hot-mopped built-up membranes. Favors LAM

You can narrow the range of practicable system selection by eliminating unsuitable systems early in the design process.

7. Specify substrate preparation in the concrete division.
8. For foundation waterproofing, favor self-adhering membranes over built-up, rubber, or PVC single-ply sheets.
9. Limit water absorption of membrane materials to two percent if practical and in no case higher than five percent (per ASTM Test Method D95).

7 Negative-side waterproofing

Negative-side waterproofing is applied to the inside (dry) subsurface space, in contrast with positive-side waterproofing, which is applied to the outside (wet) face of the subsurface building components (see Figure 5.3). Because of its ready access to the dry side of waterproofed building components, negative-side waterproofing finds its greatest use in remedial work. Several limitations – e.g., concrete's vulnerability to corrosive soil chemicals – restrict the use of negative-side systems for new construction, which is dominated by positive-side systems.

Negative-side waterproofing offers its greatest advantages on projects that would otherwise require positive-side solutions to blindside waterproofing problems. Pits, shafts, and sites where lot lines or other impediments make the outside face of a foundation wall inaccessible are examples.

Although it is effective as a waterproofing coating, the higher vapor permeability of negative-side waterproofing is a drawback in underground structures with occupancies and materials sensitive to high humidity – e.g., libraries and gymnasiums with wood floors.

Negative-side waterproofing's success in stopping leaks where other systems fail make it the system of choice in repair and remedial work. Probably 75 percent of the applications are for remediation and only 25 percent for new construction. Of the latter, five percent or more is applied on the positive-side of foundations and slabs-on-ground.

Negative-side waterproofing materials can be applied to either side of a waterproofed component. Viewed in a strictly logical light, this normally negative-side waterproofing becomes positive-side waterproofing when applied to the exterior (wet) side of a waterproofed component. Yet, when negative-side waterproofing materials (discussed in Section 7.A) are switched to the positive-side of the waterproofed component, the waterproofing industry follows a convention of still calling it negative-side waterproofing.

Superior durability and repairability give negative-side waterproofing a big advantage over positive-side waterproofing. As an integral part of the construction, negative-side waterproofing will normally last the life of a wall.

Because of positive-side waterproofing's virtual inaccessibility, it *must* last the service life of the structure. The alternative is either (a) a negative-side remedial system or (b) a very difficult and expensive positive-side re-waterproofing project. Negative-side waterproofing, however, can readily be repaired in most instances. The crystalline systems will reactivate when minor leakage occurs, and major cracking from seismic forces or settlement can be repaired by reapplication of the material.

A. Materials

Negative-side waterproofing generally comprises coatings that seal the interior surface of concrete rather than separate membranes applied to its exterior face. Membrane waterproofing is unsuitable for negative-side waterproofing because the external water pressure can easily disbond the membrane from the wall. Bentonite is also unsuitable because it requires continuous hydrostatic pressure to maintain intimate contact with the substrate.

In rare instances, mechanically adhered membranes have been used for negative-side remedial waterproofing (where existing walls are leaking) and it is necessary to restore waterproofing and provide a vapor-resistant system. These uses should be restricted to locations where hydrostatic pressure is intermittent – i.e., occurring only during periods of an unusually high water table. In such situations, the waterproofing membrane can be installed during the normal dry period. (You can specify a membrane tolerant of damp substrates instead.) After the membrane is installed, closely spaced, powder-actuated clips are driven through self-sealing butyl pads. Heavy-duty wire mesh is secured to the clips and the wall covered with gunite that is at least two inches thick. This system must be used judiciously. It is vulnerable at terminations, particularly at joints between walls and floors, where careful detailing is required to maintain watertight continuity. This will prevent the membrane from disbonding due to vapor pressure.

All contemporary negative-side waterproofing systems may be mixed with sand and troweled on or brushed or sprayed as a slurry (see Figure 7.1). The basic materials come in three generic types:

1. Crystalline coatings.
2. Cementitious coatings with metallic oxide.
3. Cementitious coatings with various densifying additives.

Crystalline coatings have to a great extent replaced metallic-oxide coatings because of their greater dependability, longer service life, and self-healing property. They are also called *chemical conversion* or *capillary coatings*.

These solutions of organic and inorganic compounds react chemically with unhydrated Portland cement and free lime in the presence of moisture. They fill capillaries and shrinkage cracks with long chain molecules that

Figure 7.1 Sprayed negative-side waterproofing coatings (left) can be troweled for smooth finish (right). (Vandex Sales & Service, Inc.)

crystallize within the concrete. One manufacturer's proprietary material includes a mixture of Portland cement, silica sand, and active proprietary chemicals. Mixed with water and applied as a cementitious coating, this material promotes a catalytic reaction that generates a non-soluble crystalline formation of dendritic (branching) fibers within the pores and capillaries in concrete. By filling the fine pores of concrete, the material impedes the infiltration of water. When the concrete cures and shrinkage cracks occur, the crystalline materials are reactivated and seal the passages.

These products were compounded in 1946 in Denmark and later in Canada around 1968. They were introduced in the United States in the 1960s, primarily as remedial systems for waterproofing failures.

Oxidized metallic cementitious coatings dominated negative-side waterproofing throughout the twentieth century. They comprise Portland cement mixed with finely divided iron or aluminum particles with ammonium or acid-based oxidation catalysts and sand in a 1:1:1 (by volume) mix. Oxidized iron particles expand several times as they corrode, thereby compacting the coating and enhancing its waterproofing properties. Where iron rust stains are objectionable, aluminum particles can be substituted in the coating. Aluminum reacts with Portland cement, releasing hydrogen-gas bubbles. Smaller than the water molecules, these hydrogen molecules disperse through the concrete matrix and make it resistant to the passage of liquid moisture.

Proprietary densified Portland cement systems are representative of the third generic type of negative-side waterproofing. Some manufacturers claim superior waterproofing qualities against high hydrostatic heads.

B. Negative versus positive

Negative-side waterproofing offers several notable advantages over positive-side waterproofing:

- easier application;
- easy leak detection;
- easy maintenance;
- high resistance to hydrostatic pressure; and
- superior concrete curing.

Ease of application makes negative-side waterproofing the first choice for remedial work where positive-side waterproofing has failed. Positive-side waterproofing is more inconvenient and costly even for new construction. Except for crystalline chemical conversion coatings, negative-side waterproofing requires little or no substrate preparation. In some situations, notably for foundations cast against rock, negative-side waterproofing is usually the most practicable solution. As ancillary benefits, negative-side waterproofing cannot be damaged by backfilling operations, and it can be inspected and repaired after backfilling is completed.

Early leak detection is an advantage because with the coating on the inside of the wall or slab-on-ground, a leak generally occurs at or near the defective waterproofing material. With positive-side waterproofing, however, the defect in the waterproofing membrane can be located some distance from the leak appearance in the interior space.

Easy maintenance results from the convenient location of negative-side waterproofing: directly at the interior surface. Defective material can be readily removed, followed by reapplication of the waterproofing coating. Reapplication of crystalline type waterproofing will generally stop subsequent leakage after initial application. However, if the concrete dries and the chemical reactions stop, it can be reactivated by the presence of water.

Resistance to high hydrostatic pressure is often an advantage, although many positive-side waterproofing systems can provide equal resistance. Crystalline and densified cementitious negative-side waterproofing coatings can satisfy the U.S. Army Corps of Engineers' CRD48 REF, which requires resistance to 200 psi hydrostatic pressure (equivalent to a 461-foot head of water). A liquid permeability in the range of 10–12 to 10–14 cm/sec is required for classification as waterproof.

Since concrete thrives in the presence of liquid moisture, which continues for months to promote hydration of the cement and its consequent strengthening and densification, negative-side waterproofing promotes **permanently ideal curing conditions**. Concrete sandwiched between water on one side and an impermeable barrier on the other is the optimum condition for healthy concrete. In the constant presence of moisture, concrete can continue gaining strength for years, though at a decreasing rate. These conditions also impede the formation of cracks from drying shrinkage, which occurs in concrete surfaces exposed to the atmosphere.

In accordance with the no-free-lunch law that rules construction as well as economics, negative-side waterproofing has notable disadvantages:

- vulnerability to substrate cracking;
- impractability for structurally framed slabs and intersecting walls;
- inability to protect against corrosive soils;
- higher vapor permeability than positive-side membranes;
- vulnerability to fastener damage;
- dependence on construction joint treatment; and
- inaccessibility when applied to concealed surfaces or blocked by piping or equipment such as switchgear.

Vulnerability to substrate cracking is the most serious liability of negative-side waterproofing. This is true despite manufacturers' claims that the chemical reaction is continuous as long as there is moisture in the concrete and that cracks are sealed as the chemicals close the pores. If applied when the concrete is fresh, the crystalline and densifying systems permit the concrete to continue curing, thus reducing cracking. Crystalline coatings can close hairline cracks – i.e. cracks up to .012 inches (0.3 mm) wide. Moisture reactivation of the sealing process gives the coatings the ability to self-seal, according to manufacturers. But no negative-side waterproofing system can bridge dynamic cracks and reseal ruptures.

Impracticability for structural slabs and intersecting walls restricts negative-side waterproofing to a fairly narrow range of applications. Readily applied to vertical surfaces or to the top side of horizontal surfaces – i.e., on slab-on-ground surfaces – negative-side waterproofing is generally unsuitable for ceiling application – i.e., the soffits of framed structural slabs. The difficulty of brushing, troweling, or even spraying these coatings onto ceilings, against the perpetual force of gravity, is generally too great. Cracks can produce catastrophic leaks, and penetration for ceiling and pipe hangers, etc., will damage the waterproofing. One manufacturer does claim to have successfully sprayed application of its negative-side product onto a ceiling.

Application of crystalline coatings to repair cracks is generally successful. Repaired cracks can, however, create a secondary leakage problem. The sealing of a relatively large leak hole can increase hydrostatic pressure relieved by the original leak. This increased hydrostatic pressure can then produce leaks at smaller holes, thereby necessitating further repair.

Negative-side waterproofing is also difficult to apply to foundation walls intersected by structural slabs or cast-in-place shear walls. Piers cast in with the foundations can readily be treated with the coating, but intersecting walls and elevated structural slabs cast into notches in the wall block the area of treatment. Negative-side waterproofing on walls cannot readily be connected to positive-side waterproofing because of the loss of continuity. Positive-side waterproofing membranes carried across elevated structurally supported slabs and down a short section of the foundation cannot be practically joined to the negative-side system.

Negative-side waterproofing **cannot protect concrete against corrosive soils** (although epoxy coatings can protect reinforcing bars); only positive-side waterproofing can protect adequately the concrete.

High vapor permeability disqualifies negative-side waterproofing where interior humidity must be closely controlled. Positive-side membrane waterproofing, which can combine a liquid barrier with an efficient vapor retarder, is required when interior humidity control is vital. Negative-side waterproofing is thus ruled out for underground spaces containing electronics, wood or carpeted floors, and other humidity-sensitive items.

Dependence on construction joint seals is a major risk. Construction (cold) joints are the Achilles Heel of cast-in-place concrete components. These appear in pressure slabs, suspended slabs, and foundation walls. Most of these joints are provided with keys to transfer shear and have continuous reinforcement. Because they are cast separately, some shrinkage will occur. Industry practice is to use either waterstops or reinjectable hoses or both. See Appendix C Waterstops.

Construction joints in pressure slabs are formed with metal or wood that is removed before the adjacent pour, leaving a smooth surface for installing waterstops. Vertical joints in foundation walls are formed in a similar fashion. However, the location of horizontal joints in foundation walls are not often predetermined and may be irregular and rough. The resulting surfaces are not suitable for preformed hydrophobic and hydrophilic waterstops, and must be ground or otherwise prepared to enable the bars to intimately and uniformly maintain contact with the concrete. Preformed plastic profiles are generally unsuitable for these irregular joints. The weakest link is where waterstops in different planes intersect. Profiled units must be welded, which is rarely successful. Preformed bars must be beveled and spliced or a gunnable waterstop must be used to form the transition. See Appendix C for a further discussion of Waterstops.

C. Application

Negative-side waterproofing coatings are applied to damp or green concrete substrates, by brushing, spraying in a slurry, or troweling (in one or more coats). Slabs-on-ground can be coated in this way or by broadcasting the dry powder by shaking over the wet concrete surface, followed by power troweling.

Slurries of dry powder and water are used in the first coat in new and remedial work and as secondary applications when the first coat requires additional treatment. These coats range in thickness from $^{1}/_{16}$ inch to $^{1}/_{8}$ inch.

Powder and water are mixed with quartz sand to a mortar consistency, and applied in corners to seal cracks; and form-tie holes, honeycombed areas, and construction joints. Exposed waterproofing may be finished with a parge coat or cement topping for decorative or protective surfacing. Parging is required on metallic oxide and cement topping for traffic areas.

The concrete substrate should be uncured or damp concrete or cement parging. Good adhesion is paramount in these systems. Early application of negative-side waterproofing does not create the risk of drying shrinkage cracks that would occur in positive-side waterproofing. As previously noted, negative-side waterproofing helps prevent premature drying of the concrete. It thus minimizes the hazards of shrinkage cracking.

Coatings adhere best when the concrete pores are opened by sandblasting or mechanical chipping. Acid etching is sometimes specified, but it has two liabilities: (1) possible exposure of reinforcing bars to chloride-ion corrosion and (2) relatively ineffective preparation of the concrete substrate. Sandblasting wet or dry or mechanical chipping is generally required to open the pores. Grinding and sanding are ineffective because they polish surfaces and inhibit penetration of the coatings. Wet and dry sandblasting are equally effective, but wet sandblasting is more ecologically acceptable. Dry sandblasting is sometimes prohibited by ordinances, because of particulate pollution. Wet sandblasting is usually more expensive. Water blasting is generally unsuitable for interior spaces.

Despite their higher cost, these more dependable substrate preparation techniques are justifiable, in keeping with the general principle of waterproofing benefit/cost analysis. Cutting construction cost is seldom, if ever, justified for even a slight increase in leakage risk. The high cost of remedial waterproofing work should be a sufficient deterrent.

Preparation for negative-side waterproofing application requires care to assure that preliminary work has been completed before application of the waterproofing coating. All cutouts for electrical boxes, floor enclosures, and other recessed items must be installed before application. Furring must be free-standing or adhered to foundation walls. Conduit and pipes must be surface-mounted. Mechanical equipment should be installed on concrete pads cast atop waterproofed slabs-on-ground to avoid penetrations from anchor bolts.

The contractor must take extreme care to avoid damage to coatings by fasteners installed after coating application. This damage can be avoided by several methods:

- eliminating fasteners and substituting adhesives
- using free-standing furring
- driving fasteners through butyl washers or pads

Because they are mixed with water and applied wet, negative-side waterproofing coatings cannot be applied in subfreezing temperatures or exposed to such temperatures before they are cured. Painting of crystalline-treated surfaces risks deterioration of the paint by continuing crystalline growth.

D. The blindside "drained-cavity" approach

A system originated in Great Britain, known as a "drained cavity," offers an essentially negative-side approach to the blindside waterproofing problem. It consists of a thick cast-in-place slab-on-ground and foundation wall, with cavities at the inside faces of both wall and slab. The wall cavity is created by a free-standing masonry wall. At the floor, a series of parallel cavities is created by a layer of inverted channel tiles, topped with a wearing course. A dampproofing membrane is recommended on the masonry wall and under the wearing course. Vertical cavities are drained to a sloped gutter. The cavity under the floor is sloped to drain and drained to a sump (see Figure 7.2).

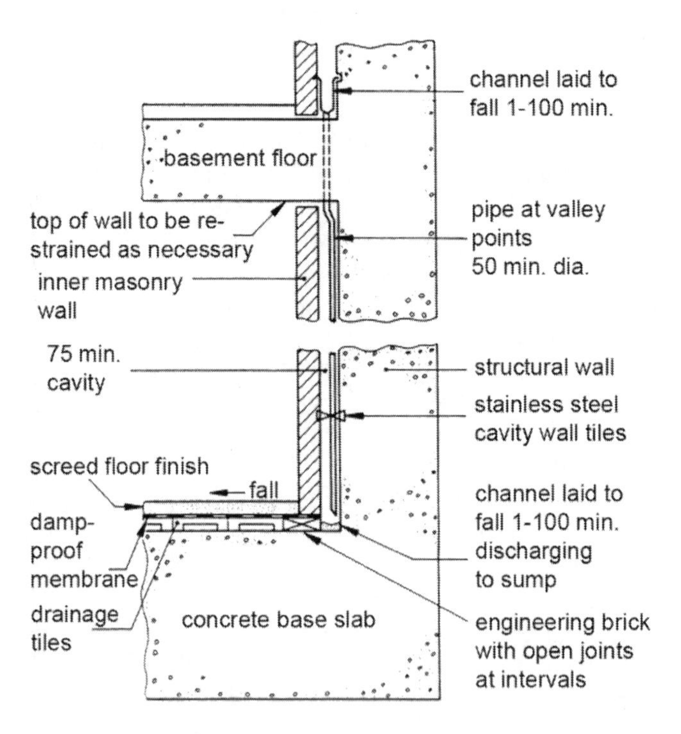

Figure 7.2 "Drained-cavity" construction poses many hazards despite its recommendation by British proponents as the "most effective and trouble-free" solution to the blind-side waterproofing problem. See text for a description of its many liabilities. (British Standards Institution) Extracts from BS 8102: 1990 are reproduced with the permission of BSI under license no. PD\1998 1273. Complete editions of the standards can be obtained from national standards bodies

According to the system's proponents, infiltrating water can be conducted to one of more sumps and discharged through pipes to daylight or to a storm drainage system. The cavities can be ventilated, either passively or mechanically, in which case the air should be exhausted to the exterior.

Despite praise from the British Standard RS8012, "Code of Practice for Protection of Structures Against Water From the Ground," (1990), as "most effective and trouble-free," its design liabilities far outweigh its assets, as it combines all the liabilities of negative-side waterproofing with some of the liabilities of positive-side waterproofing. Key liabilities are:

- Drainage water can promote organic growth, odors, and insect infestation, plus the introduction of crystallizing salts, which ultimately block drainage and obstruct flow in the wall cavity, drainage troughs, and piping.
- Exhaust-air difficulties include a requirement for introduction of make-up air at the bottom of the wall cavity, with consequent penetration of the vapor retarder, plus limitations on both gravity and mechanical ventilation.
- Sumps require a pump, possibly a back-up pump, and a generator, as power delivery may be interrupted by the flash floods that necessitate sumps.
- Serious leaks will be extremely difficult to locate, and repair will require major demolition of walls and tiles.

The system will not work where:

1. Soil contains chemical contaminants.
2. Discontinuities occur at projecting columns or shear walls; or
3. Interior humidity must be controlled.

The advantages of the drained-cavity system – adaptability to lot-line waterproofing and ability to stop water infiltration through the foundation and slab before inner walls and floor are constructed – fail to compensate for its many disadvantages.

E. Performance rating

Manufacturers of crystalline and high-density waterproofing systems rate them by water permeability, determined via US Army Corps of Engineers CRD C48-73, Permeability of Concrete. In this test concrete samples six-inch diameter, six-inch-high cylinders, coated with half-inch-thick waterproof coatings, are subjected to hydrostatic pressure of 200 psi (=461 foot head). Two treated and two untreated samples are tested. Untreated samples should exhibit a maximum 10^{-8} cm/sec. psi or less. Treated samples should not exhibit permeability exceeding 10^{12} and 10^{-14} cm/sec.

F. Checklist

1. On new construction, ban installation of negative-side waterproofing under any of the following conditions:
   a. presence of corrosive soil chemicals
   b. exposure to freeze-thaw cycling
   c. low interior-humidity requirement
2. Require manufacturer's approval of applicator.
3. Require pre-installation inspection of substrate surfaces.
4. Cite surface preparation requirements.
5. Require waterstops for negative-side waterproofing.

8 Blindside waterproofing

Blindside waterproofing is a system in which a below-grade waterproofing membrane is temporarily attached to the soil retention system facing the excavation prior to casting the concrete foundation against it. It is required where the exterior faces of foundation walls will be inaccessible. A common situation dictating blindside waterproofing is the proximity of adjacent property lines or other abutting structures, which preclude excavation outside the foundation walls (see Figure 8.1).

Figure 8.1 A two ply modified bitumen waterproofing membrane is shown being installed under the pressure slab and carried up the boarded-out furring on this law school basement. Note the well points

Strictly defined, blindside waterproofing is positive-side waterproofing, but a blindside waterproofing problem is sometimes solved by negative-side or integral waterproofing. Waterproofing under pressure slabs on grade is also a form of blindside waterproofing.

This chapter discusses the pros and cons of the various types of blindside waterproofing membranes that are currently marketed in the U.S. It does not include products manufactured to be used on plazas and similar locations where the waterproofing system is not subjected to hydrostatic pressure. It also excludes crystalline and similar coatings applied to the negative-side of foundations and other additives to cast-in-place concrete.

A. Historical background

Below-grade occupied spaces were once provided to house building heating systems and served as storage areas. They were rarely deeper than one story. Since World War II, deeper basements have been required for below-grade automobile parking and air-conditioning equipment.[1]

Buildings in large cities were being built out to lot lines and fronted on streets crowded with below-ground utilities. Thus, applying waterproofing to the outboard side of foundations became impractical.

Today's foundations are typically reinforced concrete. Prior to the turn of the nineteenth century, foundations were primarily constructed of stone; several feet thick and frequently parged with cement on the inboard side. The water resistance of the parging was increased by adding iron filings and an oxidizing catalyst that caused the iron to corrode and swell, compacting the parging.

Limited moisture infiltration in these stone foundations often occurred, but was not considered critical since the equipment located in the basement could tolerate dampness. However, when basements were constructed in areas with high water tables, more water-resistant construction was required to satisfy the demands of moisture-sensitive occupancies. The choices were to construct a watertight basement with walls and slabs sufficiently strong to resist hydrostatic pressure or to provide a drainage system of sumps and pumps in order to reduce water pressure. The proponents of the drainage system contended that the "interest on the increased cost of constructing a hydrostatic pressure-resistant basement was greater than the cost of providing sumps and operating the pumps."

Chemicals were frequently added to concrete to densify it by filling its voids. These additives consisted of finely ground sand, colloidal clays or hydrated lime. Some additives, such as stearates and oil, were intended to repel water. In 1910, the *Engineering News* reported that "oil in the amount of 10% by weight of the cement gives very satisfactory results" and is economical "since oil costs about 6 to 7 cents per gallon or about 60 to 70 cents more per cubic yard of concrete."

The use of additives was not universally accepted. Kidder-Parker, in its

1945 Handbook,[2] recounts a report published by ASTM Committee D08, circa 1927, regarding the permeability of concrete and methods used to waterproof it. The committee report discusses the results of laboratory testing and information obtained from the field. It evaluates the "Addition of Foreign Substances" and "External Treatments." The committee concluded that additives to concrete were of doubtful benefit, whereas protective coatings, both bituminous applied to the exterior faces of the concrete and cementitious applied to the interior face, had proven to be efficacious.

At the turn of the twentieth century, popular blindside waterproofing applications on zero-lot-line basements consisted of erecting drainage tile against a soil retaining brick wall, then covering them with bricks dipped in hot coal tar pitch. Asphalt bricks were also used and dipped in hot (unblown) asphalt. Alternately, multiple layers of burlap or felt swabbed with hot pitch or asphalt would be applied over the tile and covered with bricks. The concrete foundation wall was then cast against it. Boilers, which were located in the basement, kept the foundation walls warm enough to keep the coal tar pitch soft and enable it to reseal when shifting soils caused seams to open.

Good workmanship was critical to the successful waterproofing envelope, but perfection not always achieved. Consequently, prudent designers did not solely rely on the waterproofing system but installed a drainage system under the waterproofed slab that conducted infiltrating water to a sump, which pumped it to a sewer. Today, few municipal sewage treatment plants will accept the discharge of subsurface water, or will only accept a limited amount.

B. Blindside waterproofing membranes

Generally, blindside membranes can be divided into two types:

1. Attached.
2. Non-attached.

Attached types are intended to bond mechanically or adhesively to the concrete after it is cast against them. Non-attached types are faced with a granular bentonite compound or a hydrophilic polymer, which is intended to swell when in contact with water, and form an impermeable watertight gel between the soil and the concrete or a loose-laid thermoplastic sheet.

Before 2000, only a handful of manufacturers produced membranes specifically for blindside waterproofing. Most of these products were either bentonite clay systems or one-ply or built-up membranes designed to adhere to the concrete cast against them. Adhesion was obtained chemically or by concrete mechanically engaging fibers.

The use of blindside waterproofing increased as more buildings were constructed in heavily populated areas. That precluded the luxury of extending the soil retention system sufficiently beyond the foundation to permit application on the positive side of foundation walls. The growing market prompted more waterproofing manufacturers to develop new products. These include bentonite clay-based compounds and hydrophilic polymers laminated to ethylene interpolymer (EIP), HDPE and geotextiles, a self-adhering styrene butadiene styrene (SBS) laminated to a polyester fleece, an adhesive surfaced HDPE, a polymer modified asphalt emulsion and a butyl alloy laminated to thermoplastic polyolefin (TPO). Modern products include:

- a bentonite or mostly bentonite sheet applied to the soil retention system that will create a water-impermeable gel between it and the concrete that is cast against it.
- a water-impermeable sheet applied to the soil retention system that will reattach itself to concrete cast against it and thus prevent leak water migration.
- a sheet applied to the soil retention system that is water impermeable and contains bentonite or hydrophilic polymer based compounds and will prevent leak water from reaching the concrete. Some sheets are designed to reattach to the concrete cast against them.
- a multiple-ply, cold-applied two-component polymer modified asphalt emulsion.

All these systems are intended to facilitate leak detection by limiting the migration of leak water between the membrane and the concrete.

1. Bentonite-based membranes (also see Chapter 12)

Older bentonite clay systems, such as sprayed and troweled-on applications and clay filled kraft paperboard panels, have generally been abandoned by manufacturers in favor of bentonite encapsulated in geotextiles or bentonite laminated to thermoplastic sheets.

The new products do not fail as their predecessors did when the bentonite was washed away by flowing water. However, they do fail when the soil retention assembly develops voids and does not provide the requisite confinement.

2. Bentonite/thermoplastic laminated membranes

The thermoplastic laminated products are intended to correct bentonite clay erosion by introducing a sheet membrane as the first line of defense and utilizing the gel-forming bentonite bonded to it to prevent water that infiltrates the seams from coming in contact with the concrete.

The thermoplastic sheets alone are unsatisfactory blindside waterproofing membranes because they lack the ability to prevent leaking water from migrating laterally between the concrete and the membrane. The bentonite accomplishes this, but only when sufficient pressure is permanently maintained between the substrate and the membrane to confine the bentonite gel. Karim Allana reported on an HDPE/bentonite failure that resulted from moisture that deformed wood lagging.[4]

Newer granular bentonite compounds laminated to HDPE thermoplastic sheets are a potential solution. One has an additional geotextile facing the concrete that is intended to resist hydration on vertically applied installations, but not bond to concrete.

3. Bentonite/geotextile membranes

These products encapsulate bentonite between layers of woven and non-woven polypropylene. The bentonite is intended to swell when hydrated, and one product is faced with a geotextile that is claimed to form a mechanical bond with the concrete cast against it (see Figure 8.2). Also, see Chapter 12.

Figure 8.2 Bentonite encapsulated in geotextiles is shown installed under a pressure slab in a Philadelphia stadium. The membrane is installed over tamped gravel and the seams are lapped. They are secured with staples or fasteners

Excluding products faced with geotextiles, the bentonite membranes rely on the principle that water-tightness can be obtained by maintaining permanent compression between the soil retention system and the concrete foundation. This enables the swelling clay to develop sufficient pressure to prevent water from reaching the concrete and migrating laterally. The effectiveness of the soil retention system in providing solid, void-free backing is critical to its success. In practice, settlement, intermittent pressures, corrosion, and rot can combine to erode the structural integrity of the substrate. That may reduce the confining pressure below that is required to ensure watertightness. Gibbons and Towle[5] note that the West Coast practice of using shotcrete as a soil retention system "does not provide the necessary confining pressure to allow bentonite platelets to...create a waterproof layer."

4. Bituminous membranes (also see Chapter 9)

The use of multiple-ply coal tar pitch membranes has been so restricted by VOC regulations that they are rarely used in new construction. They have been replaced by a two-ply chloroprene modified asphalt membrane and an SBS modified asphalt sheet laminated to polyester fleece or non-woven fabric. The older chloroprene modified membrane is reported to bond to the concrete when it is softened by the heat of hydration of the cement when the concrete is cast against it. The newer membranes are also intended to mechanically bond to the concrete cast against it. Adhesion is also obtained by the lateral pressure of the wet concrete. One manufacturer markets a spray-applied two-component modified asphalt emulsion that it claims will chemically bond to concrete cast against it. The membrane is also marketed as a gas vapor barrier system.

The older two-ply membrane has a 45 year track record, whereas the newer SBS membranes have yet to establish one. The asphalt emulsion has a somewhat limited resistance to hydrostatic pressure compared to the other two.

5. Thermoplastic and HDPE/Adhesive surfaced membranes (also see Chapter 10)

One of the oldest single-ply blindside waterproofing membranes consists of an HDPE sheet surfaced with an adhesive. A recently introduced system consists of a TPO membrane surfaced with a butyl adhesive. Both membranes are intended to chemically bond to the concrete cast against them aided by the pressure of the wet concrete. Seams are sealed with pressure sensitive tape.

The HDPE/adhesive sheet manufacturer does not recommend it for use in a blindside application where the adhesive face will receive shotcrete. They assert that the loss of pressure from wet concrete and shadowing by rebars inhibits adhesion.

The HDPE sheet is thick and does not easily conform to changes in plane. The manufacturer recommends using a self-adhering rubberized asphalt sheet and a two-component, trowel-applied asphalt modified urethane and tapes. This transition has proven to be the most vulnerable part of the system and must be carefully designed and installed (see Figure 8.3).

The TPO/Butyl sheet is a self-adhered 22 and 30 mil TPO, which is more flexible. The seams are lapped and sealed with factory-applied adhesive. Whether this seam performs satisfactorily when the 30-mil sheet is turned up the foundation and concrete is cast against it has not been proven yet.

C. Seams

As with all sheet membranes, seams are the Achilles Heel where water infiltration is most likely to occur. Movement of the substrate can open so-called bentonite pressure seams that are simply lapped or mechanically fastened and depend on soil compression to keep them watertight. Taped seams improve the integrity of the seam since they are better able to resist the shrinkage that is common with HDPE. Although more expensive, heat-fused seams offer by far the greatest resistance to seam-failure caused by differential movement between the concrete and the substrate.

Figure 8.3 This HDPE membrane is being installed under the pressure slab and is turned up behind the foundation wall. Seams are being taped

The location where seams are probably most critical is the transition from horizontal to vertical and the vertical reentrant angle. Thermoplastic sheets are usually thick and can be quite stiff in colder climates. They are often thicker for use under pressure slabs than sheets applied to the lagging. At the horizontal/vertical reentrant angle, most manufacturers' details suggest that the horizontal membrane be turned up the wall a specific distance and lapped by the vertically applied membrane. This often results in tenting because the sheet is too stiff to conform to the right angle. The adhered seam is also turned upward. When the concrete foundation wall is cast against the coved corner, the seams are susceptible to rupture or disbonding. The problem is exacerbated at the transitions where the slab meets the interior and exterior corners of the foundation, columns, and pile caps. With sheet thicknesses that exceeds 60 mils, good tight corners are virtually impossible to fabricate in one piece.

Single-ply waterproofing membranes lack redundancy. When they are adhesively bonded (as opposed to heat fused) the seams are subject to stress from shrinkage and displacement from the weight of the concrete. Prudent designers will specify that the seams be taped which provides a second adhesive seam. Where foundations are deep and hydrostatic pressure is high, additionally taping the back of the seam will provide help ensure a leakproof seam (see Figure 8.4).

Figure 8.4 The back of the seams of this blindside HDPE membrane have been reinforced with tape. The seams facing the concrete foundation wall will also be taped

Joints of thermoplastic sheets that are welded or that use a combination of more flexible sheets and tapes and a liquid component usually perform better than those that rely on adhesives or bentonite pressure laps.

D. Attachment to substrate

A basic premise of blindside waterproofing is that the membrane must be installed securely, albeit temporarily to the soil retention system. It should resist displacement from sagging wet concrete and be capable of spanning small voids, step-offs, and other surface irregularities without rupturing or disbanding seams. It should not be secured so tenaciously that displacement of the soil retention system tears it away from the cured concrete foundation. Moreover, it should provide uniform support free of voids that can localize pressures that rupture the membrane. Seismic events, corrosion, erosion warping, and decay must be considered.

Soil retention systems usually consist of lagging or shotcrete but also may include sheet piling and secant piles. Lagging is usually installed with one-inch or larger joints between timbers and must be overlaid to form a smooth solid surface to receive the blindside membrane. This is the role of plywood sheathing, drainage composites, protection boards, and rigid insulation (see Figure 8.5).

Figure 8.5 The soil retention soldier pile and lagging system is shown being boarded with a layer of one pcf density EPS insulation covered with number 15 felt and plywood. The low-density EPS will fail in shear when differential movement between the SOE and foundation wall occurs

Assume that the attachment of the membrane to the soil retention system will be temporary and that the fasteners will corrode and/or the substrate into which they are driven will rot away or disintegrate. Eventually, the lagging will no longer provide a uniform, structurally sound support. If the retention system shifts, rots away, twists, cups, or disintegrates, the membrane may be vertically or laterally displaced or the resultant voids will fail to provide the requisite pressure required by bentonite and its watertight integrity will be threatened.

All three systems rely on watertight seams. Sheets containing bentonite depend on the initial and continuing integrity of the soil retention system to maintain void-free solid surfaces that will be capable of resisting water infiltration.

Blindside membranes that depend on adhesion to concrete share this problem of unstable substrates, but not to the same degree. Shifting, settlement or similar lateral movement of the soil retention system can shear the tenuous bond between the membrane and the concrete, tear it at fasteners, or open seams between the sheets. This can often be minimized by introducing several layers of materials between the membrane and the lagging. Common components are plywood, protection boards, drainage composites, and low-density expanded polystyrene. Components can be used separately or in combinations. Differential movement between the membrane and lagging can be absorbed if these components are adhered, rather than mechanically fastened or if low-density expanded polystyrene is allowed to shear internally. However, fasteners may be required to resist the shear forces that result from placing concrete.

Adhered and non-adhered membranes must be attached to a relatively smooth and uniform substrate. Voids, step-offs, and other surface irregularities must be corrected to avoid pressure from the wet concrete from rupturing the membrane or disbonding the seams. Secant piles can be shaved down to the flanges of the steel piles. But the other types of support of excavation (SOE) walls must be boarded out with plywood, drainage composites or insulation, or combinations thereof to offer a satisfactory surface. Plywood must be secured with fasteners that will not back off and penetrate the membrane.

E. Membranes under pressure slabs

Membranes for use under pressure slabs are similar to those used for blindside waterproofing on foundations; envision a blindside waterproofed foundation rotated 90 degrees. They are intended to bond to the underside of the pressure slab that is mechanically or adhesively cast over them, or to swell in contact with water to form a water-impermeable gel.

The membranes are installed over a well compacted gravel subgrade or an unreinforced concrete mud slab. The gravel must be free of voids or pockets that would permit the membrane to bridge. The mud slab must also be free of voids, ridges or surface irregularities that would not provide uniform support. Neither the mud slab nor compacted gravel is intended to provide support for the membrane for the life of the building. The gravel will eventually settle or be washed away and the mud slab will disintegrate. When this occurs, the membrane must remain firmly adhered to the pressure slab or the bentonite must remain in compression. In this respect, the bonded membrane offers superior water resistance.

However, the ability to provide a satisfactory mechanical or adhesive bond can be compromised during the normal course of construction. Sheets that depend on mechanically engaging concrete with geotextile fibers can have those fibers compressed by construction traffic and material storage to the extent that they can no longer provide a satisfactory bond. Sheets that depend on chemical adhesion can lose their adhesion when coated with a film of dirt that concentrates in puddles over the surface.

Both problems can often be avoided by casting a four to six-inch layer of concrete over the membrane as soon as possible after each section is completed. This will protect the membrane, and its surface can then be raked to bond to the pressure slab. The procedure has the added advantage that reinforcing chairs, pipe supports, and the like will be supported on the concrete fill, rather than directly on the membrane.

F. Checklist

Failures in blindside waterproofing applications may not be avoided by using one of the products introduced to the market in recent years, but you can take precautions to reduce the risks and provide a long-term watertight basement.

1. Select products from companies with experience in manufacturing and installing blindside waterproofing systems.
2. Involve the manufacturer's technical experts early and throughout the initial phases of the installation.
3. Prepare exhaustively detailed construction and shop drawings that address penetrations, plane changes, tie-backs, walers, raker's footblocks, pile caps, construction joints, and allied issues. Don't rely on stock details.
4. Conduct preconstruction meetings with the excavation contractor to explore all the means and methods of soil retention. This includes the location of well points. They are never indicated on drawings, but if they are located too close to the foundation walls, proper flashing becomes extremely difficult.
5. Provide hydrophilic rubber, with or without bentonite waterstops, at all construction joints in foundation walls and pressure slabs on grade. Consider adding re-injectable hoses where pressures are high and occupancies are sensitive to moisture.

6. Advise your client that some leaking cannot always be prevented but that there are established means and methods to stop water infiltration, should it occur.
7. Use a concrete protection slab over adhesive coated membranes under pressure slabs.
8. Use a mud slab under pressure slabs rather than using compacted gravel.
9. Carefully detail sheet plane transitions. Use cants or an assembly of sheets, tape, and the liquid component of the system rather than relying on the manufacturer's details.
10. Ensure that the waterproofing contractor's superintendent inspects the membrane prior to casting each lift of concrete because lathers that erect reinforcing and form spreaders often damage in-place waterproofing.

Notes

1 Following super storm Sandy and hurricane Katrina, many municipalities on the coasts raised their flood stages and together with insurance companies, required that mechanical equipment and switchgear be relocated to higher floors, leaving the basement for tanks and parking.

2 *Kidder-Parker, Architect's and Builders' Handbook*, 18th Edition, 1945, John Wiley & Sons, New York, NY.

3 Properties of membranes have been obtained from the latest literature published by: W.R. Grace, Carlisle Coatings & Waterproofing, Tremco, Soprema, Cetco, W.R. Meadows and Laurenco.

4 Allana, Karim, Bentonite Composite Waterproofing System Below Grade Applications Failures and Solutions, Sealant, Waterproofing and Restoration Institute, 2010 Winter Technical Meeting, South Beach Miami, FL Feb. 21–24 2010.

5 Gibbons, Daniel G and Towle, Jason L., Waterproofing Below-Grade Shotcrete Walls, *The Construction Specifier*, March, 2009.

9 Built-up bituminous waterproofing membranes

Built-up bituminous waterproofing membranes have been fading in popularity for several decades, supplanted by a plethora of new materials and systems. Yet, you still need to know about them for several reasons:

1. They were widely used for more than a century, and failures encountered with them contain lessons applicable to newer materials. For instance, slippage of built-up membranes on vertical surfaces, a result of gravity combined with aggravating frictional forces exerted by settling backfill, can pose a similar problem for other wall-waterproofing membranes.
2. Remedial work involving an existing built-up membrane requires knowledge of these materials, even when the remedial material is modified bitumen or another contemporary membrane. A vast number of existing waterproofing projects have these materials in place.
3. Some clients still insist on using traditional built-up membranes, despite the problems and obstacles they pose.

A. Description, background, benefits, and disadvantages

Built-up waterproofing membranes are semi-flexible laminates, comprising alternating layers of normally hot-applied bitumen and felt (or fabric) reinforcement. As positive-side waterproofing components, they are fully adhered to walls and below-grade structural slabs.

Bitumen, the most important element, serves a dual function as waterproofing agent and adhesive. Reinforcing felts and fabrics function like veneers in a sheet of plywood. They provide at least 90 percent of the membrane's tensile strength. Reinforcement also stabilizes the membrane, acting as sort of a matrix. By isolating the multiple layers of waterproofing bitumen during installation, reinforcement plies promote uniform thickness in the adhesive films that bond the membrane into a structural unit.

As their chief advantage over other types of waterproofing membranes, built-up membranes exploit a property known as redundancy. It denotes the built-up membrane's multiple defense lines, with each of the normally three plies of reinforcement adding an extra plane of waterproofing bitumen. Redundancy accounts for the extraordinary durability of well constructed, built-up membranes.

Along with their counterpart built-up roofing (BUR) membranes on low-sloped roofs, built-up membranes are the oldest waterproofing system, introduced in the nineteenth century. They are the evolutionary progeny of the archetypal waterproofing project, the Hanging Gardens of Babylon, which exploited the waterproofing property of bitumen more than 25 centuries ago. Many competing products – prefabricated modified bitumens, prefabricated elastomeric and thermoplastic sheets, and liquid-applied coatings – have steadily gained market share. Built-up membranes share a minimal and decreasing part of the waterproofing market because of the relentless march of air-pollution regulations. Built-up coal tar pitch membranes have been almost completely withdrawn from the market as a result of VOC requirements.

On projects complicated by the aforementioned factors, built-up membranes become impracticable; their advantages outweighed by their disadvantages. Foremost among these disadvantages is the difficulty of flashing built-up waterproofing systems. Complicated flashing problems call for an alternative system – modified bitumen, prefabricated, single-ply sheets, or, in extreme instances, even liquid-applied membranes. Built-up waterproofing membranes and liquid-applied membranes are at opposite ends of the spectrum. For large expanses of uncomplicated waterproofing, where the objective is to provide dependable, durable waterproofing, hot-applied bitumen is among the prime candidates unless environmental conditions preclude heating bitumen. Where the project features complicated geometric contours and numerous penetrations, a liquid-applied system may be best. For intermediate situations, prefabricated single-ply sheets probably offer the best compromise.

B. Materials

Waterproofing bitumens are the same basic materials used for built-up roof membranes but restricted to unblown asphalts in the lower viscosity range. As noted above, coal tar pitch is no longer a factor in the market due to VOC regulations.

Petroleum asphalt (differentiated from natural asphalt) is the dense "bottom-of-the-barrel" residue left after petroleum distillation drives off the lighter hydrocarbon compounds used in gasoline, jet fuel, and a huge number of other products. The asphalt content of crude petroleum varies from none to more than half.

For waterproofing, unlike roofing, water repellence is the paramount performance criterion, with nothing else even a close second. In its protected, better insulated locations, waterproofing asphalt can sacrifice other desirable properties to this overriding requirement. It is normally unblown asphalt, unlike asphalt subjected to the blowing process used to produce the harder, less viscous asphalt used for roofing.

Among the liabilities of blown asphalts is deterioration from contact with soil contaminants, especially hydroxyls. These vulnerable blown asphalts may hydrolyze in the presence of water contaminant chemicals. This corrosive process makes the asphalt water-soluble. The surface may erode with the rise and fall of water tables or the flow of underground water. Built-up waterproofing membranes made from these asphalts can suffer severely shortened service lives.

Waterproofing, like roofing, when applied on structural slabs, requires a slope to drain. Recommended minimum slope is one percent ($^{1}/_{8}$ inch per foot), but settlement, creep, and concrete casting tolerances can level the surface. Therefore, to maintain positive slope, two percent is recommended. The maximum slope depends upon the waterproofing materials, primarily on bitumen viscosity.

Waterproofing bitumens require high penetration and a low softening point, an index of bitumen's tendency to flow when subjected to rising temperature. Tested via ASTM standards, these properties are indexes of the bitumen's ductility or toughness, important properties of membranes subjected to the dynamic stresses accompanying backfilling and placing of overburden.

Reinforcement stabilizes waterproofing bitumens and allows the membranes to be built-up to the specific quantities required to resist hydrostatic pressure. Reinforcing felts and fabrics come in a broader spectrum of materials for waterproofing than for roofing. Felts are glass fiber or polyester.

Organic felts contain cellulose fibers from shredded wood and felted papers. These felts are vulnerable to fungus rot, which weakens and eventually destroys them. Organic felts are not recommended for asphalt built-up membranes.

Fiberglass felts, the predominant felts in both roofing and waterproofing, entered the U.S. market in the late 1940s. Long, glass filaments drawn from molten glass streams through tiny orifices made of precious metals. The glass filaments are usually bound with binders – phenol formaldehyde,

Table 9.1 Table of recommended softening points for waterproofing bitumens. Type I asphalt and Type II coal tar pitch have self-sealing characteristics

Product	*ASTM*	*Softening point °F*
Asphalt	D-449	
	Type I Type II Type III	115–145 145–170 180–200
Tar pitch	D-450	
	Type II	120–140

urea-formaldehyde, or acrylic resin – and coated with hard-grade, blown asphalt. In the cross-machine (i.e., transverse) direction, glass fiber felts have tensile strengths one-and-a-half to four times greater than organic felts. Unlike organic felts, glass fiber felts are water resistant, and thus immune to deterioration from fungus rot.

Fabrics of cotton and jute (burlap), coated with asphalt or coal tar, were traditionally used to reinforce built-up waterproofing membranes, either alone or in combination with organic felts. Like organic felts, which have been largely replaced with glass felts, cotton and jute have mostly been replaced with woven-glass fabrics. Cotton and jute are now mostly used with cold bitumens. Jute, a very coarse fabric, can hold more bituminous material and is thus favored for vertical surfaces and flashing. Glass fabrics are coated with asphalt or resin. They possess superior strength, are rot-proof, and conform to irregular surfaces. They are used mainly for flashing and for reinforcing corners, and also with glass and organic felts as membrane reinforcement.

In summary, you should avoid all other types of organic felts and organic fabrics for waterproofing membranes. For corner-reinforcing and flashing, use woven-glass fabric or polyester felts. For asphalt built-up membranes, specify polyester or glass fiber felts (ASTM D2178, Type IV or Type VI), which perform well.

C. Specifications

Specifications for built-up waterproofing membranes range from a minimum of three (two, in rare cases) to a maximum of six plies. Since the 1930s, manufacturers have published tables listing the number of plies recommended for different hydrostatic pressures (measured in feet of water, or head (=62.4 psf/ft). Some authorities believe you should specify an absolute minimum of two plies (generally increased in practice to three) for heads less than three feet, and up to six plies for heads of 50 to 100 feet. However, correlating the number of plies with hydrostatic pressure has no scientific basis. Three plies of inorganic felt are normally satisfactory, but four plies are recommended.

For equal material and construction quality, additional plies increase a waterproofing membrane's water repellence and durability. For the minimal saving of reducing the number of plies, the addition of virtually any incremental risk is probably unjustified. Saving the cost of onsite inspection is even less justifiable on a life-cycle costing basis since poor application practice is a major source of waterproofing membrane failure and doubtless poses a greater incremental leakage risk than the reduction of one ply of reinforcing.

Unlike built-up roof membranes, where shingling of felts is almost universal, built-up waterproofing membranes are usually "phased." In phased construction, the waterproofing contractor applies alternating layers of felt (or fabric) in separate "ply-on-ply" or "one-on-one" patterns instead of overlapping the felts, shingle-style, in a continuous operation that often completes the entire membrane. The separate operations of phased application provide superior waterproofing because they isolate each bitumen layer from the adjacent layer, creating, in effect, a multi-ply series of waterproofing layers (see Figure 9.1).

Shingled felts, in contrast, can eventually wick moisture from an exposed felt edge diagonally down from the top of the built-up membrane to a base sheet, or, in a totally shingled membrane, through the entire cross section to the substrate. Defective application resulting in a fishmouth or other lap defect can thus open a direct leakage path through the entire membrane cross section. In a "one-on-one" patterned membrane, a similar surface defect is far less likely to result in leakage.

As a compromise between the shingled and phased application, waterproofing experts have devised a combination of phased and shingled application for membranes containing five (or six) plies. Application of three shingled plies is followed by two (or three) shingled plies. This procedure creates a continuous waterproofing film between the two multi-ply installations, meanwhile giving the applicator the convenience of shingling. Laying the top two plies perpendicular to the lower three plies also produces a membrane with more nearly isotropic strength and strain properties (a particular advantage with organic felts).

Ply-on-ply construction is, however, the norm for built-up waterproofing membranes through the entire membrane cross section, as it promises better waterproofing quality. This procedure does require greater care in felt-laying operations: ensuring that the felt is laid within several seconds after the hot bitumen is deposited. It also limits or eliminates work in poor weather, especially in strong wind conditions. Even at relatively high ambient temperature, increases in wind velocity exponentially accelerate hot bitumen's cooling rate.

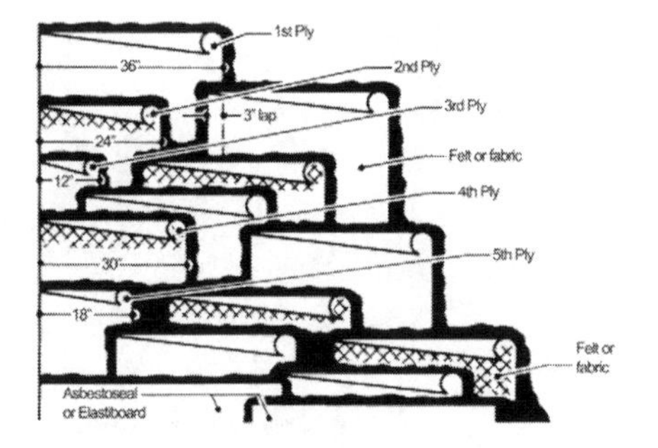

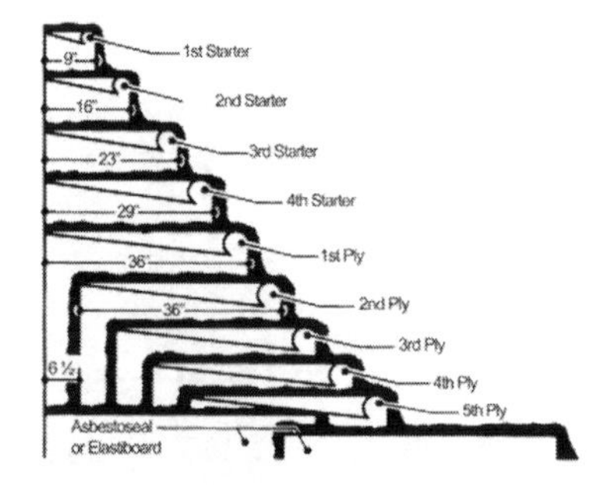

Figure 9.1 Diagrams of shingled and phased applications. "One and One" provides a continuous film of bitumen. The direction and materials of alternate layers can be changed

Combining different reinforcement materials to exploit their complementary properties offers several benefits in built-up waterproofing membranes. Alternating glass fiber felts with woven-glass fabric alleviates the threat of interply mopping voids caused by glass fiber felts "memory" – i.e., its tendency to retain kinks and other irregularities instead of relaxing like woven-glass fabric after installation.

D. Vertical surfaces

As previously indicated, application of built-up membranes to vertical surfaces is much more difficult than application to horizontal substrates. Instead of retaining stable positions after the adhesive bitumen cools, felts can slowly slide down vertical surfaces with the flow of the less viscous waterproofing bitumens. Asphalt built-up membranes require anchorage, preferably with roofing nails driven into horizontal treated wood nailer strips, but at least with concrete nails into a concrete masonry substrate (see Figure 9.2).

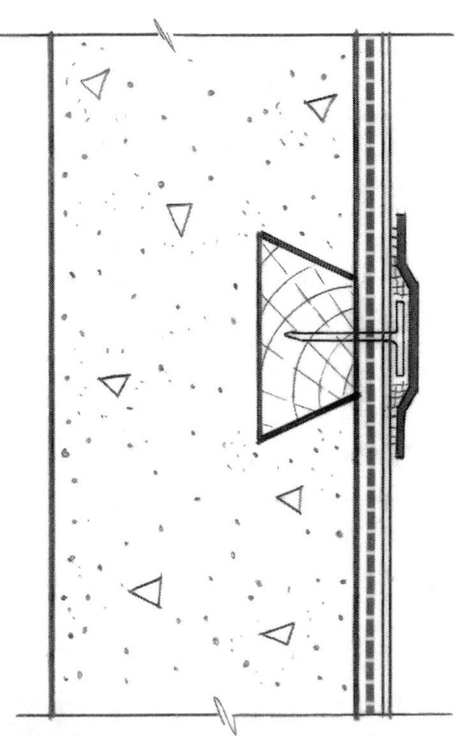

Figure 9.2 Vertical cross-section through foundation wall shows top of membrane and protection board anchored to foundation wall with a capped masonry nail. It is driven by a horizontally oriented wood nailer at 12-inch spacing, two-to-five inches below top of sheet. A four-inch-square piece of protection board set in roofing cement shields masonry nail cap

All built-up membranes are anchored at the tops of foundation walls. This top anchorage may be sufficient for asphalt membranes smaller than 10 feet. Anchorage is obtained by backnailing at a minimum vertical spacing of 10 feet.

In vertical applications the felts are usually laid vertically. They are cut into six- or eight-foot lengths and either shingled or laid ply-on-ply. Application may proceed full height, for a length of the wall. The membrane is then covered with protection board and backfilled. Backfilling each lift eliminates the need to install scaffolds, with the application proceeding from the bottom up. Because mops cannot carry as much bitumen to vertical surfaces, mopping twice is often necessary to attain specified interply mopping weight. Note that built-up membranes require a cant at reentrant angles. They are also reinforced at interior and exterior corners, both vertical and horizontal (see Figure 9.3).

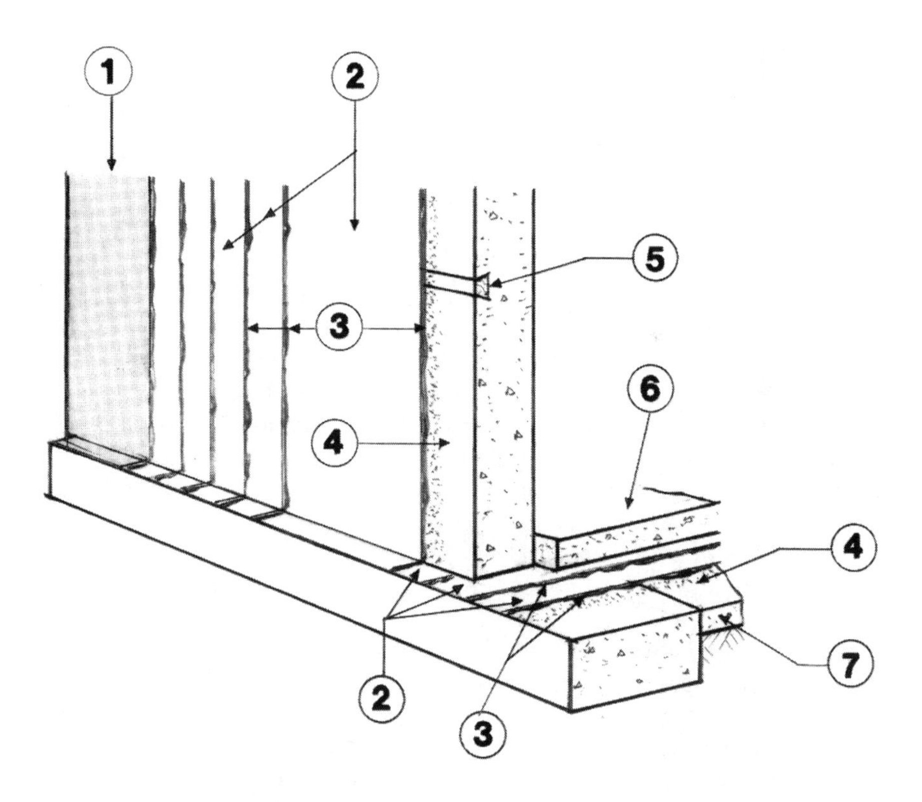

Figure 9.3 Typical application of plies on foundation wall: (1) Protection board, (2) felt or fabric plies, (3) bitumen, (4) primer, (5) nailer, (6) structural slab, (7) mud mat

All built-up membranes, for both slabs and walls, require priming of substrates to provide a tacky surface assuring continuous, dependable adhesion. Priming of masonry and concrete surfaces absorbs dust and other contaminants that impair proper adhesion and create voids that expose the membrane to punctures or progressive disbondment.

E. Causes of built-up membrane failures

In their normal subsurface locations, shielded from solar radiation, extreme temperature changes, and wind forces attacking roof membranes; built-up waterproofing membranes are virtually immune to several maladies that afflict built-up roof membranes, notably blistering and ridging. They do, however, share in built-up roof membranes' vulnerability to:

- slippage (on vertical surfaces);
- rotting of reinforcing felts;
- punctures;
- splitting; and
- defective flashing.

Slippage on vertical surfaces is a major problem for built-up waterproofing membranes. As previously noted, built-up membranes require backnailing anchorage on vertical surfaces. Slippage failures can leave partial areas of foundation wall unprotected as the membrane slowly descends.

Avoiding slippage failures is not difficult. It requires backnailing built-up membranes on vertical surfaces. Horizontal wood nailers should be cast into the exterior concrete wall face at a maximum vertical distance of 10 feet for asphalt membranes. All membranes require backnailing at the top (see Figure 9.2).

Backfill settlement is an aggravating factor promoting slippage. To the gravitational slippage force, settlement adds the frictional force of downward-moving soil mass in contact with protection boards. See expanded discussion of backfill settlement in Chapter 14.C. "Foundation wall failures."

Rotting of organic felts rarely a problem as their use has drastically declined. Organic fibers are notoriously susceptible to fungus rot promoted by moist conditions.

Punctures and abrasions can destroy the waterproofing ability of built-up membranes. The contractor must install protection boards immediately after completion of membrane application to prevent damage. Membranes on vertical surfaces can be damaged by backfill operations, even with protection boards in place, if the backfill includes sharp-edged rocks or the bulldozer or grader operator slams a steel blade against the wall. Unprotected membranes on slab-on-ground and structural slabs (lacking protection boards) can be damaged by reinforcing bar supports, the bars themselves, edges of reinforcement mesh, electrical conduit, pipes, foot traffic, concrete buggies, vibrators, rakes, spades, and other puncture-threatening hazards associated with construction.

Built-up membranes (like other types) are also vulnerable to rupture from hydrostatic pressure exerted against an irregular substrate (e.g., a honeycombed concrete surface). This pressure can exert a combination of shearing and flexural stresses and ultimately rupture the membrane. Materials with greater elasticity – notably modified bitumens, PVC, and synthetic rubber – have a great advantage over built-up membranes in resisting such stresses.

Lack of adhesion to the substrate; resulting from moisture, curing agents, and omission or defective application of primer can also promote puncture of built-up membranes (as well as other types.)

Splitting is a lesser problem for built-up waterproofing membranes than for built-up roof membranes because waterproofing membranes are not subjected to the extreme temperature dimensional changes experienced by exposed roof membranes. Splitting in waterproofing membranes is limited to the propagation of tensile stresses from shrinkage-caused cracking of concrete or masonry substrates, or failure to reinforce corners and dynamic construction joints.

Defective flashing results from use of felts instead of more flexible fabrics at external corners. Erroneous use of felts also makes membranes vulnerable to puncture or other mechanical damage. It may also open a leakage path, where stiff glass fiber felts, in particular, fail to close an overlapped joint.

F. Checklist

Provide minimum 1/8-inch (preferably 1/4-inch) slope (1 percent) in structural decks.

1. Slope decks away from expansion joints and rising walls.
2. Avoid organic felts and asphalt where they will be in constant contact with water.
3. Use fabrics for plane transitions. Glass felts, in particular, cannot be carried around corners.
4. Detail cants and bevels at corners.
5. Specify one-on-one, phased application rather than shingling.

10 Single-ply waterproofing membranes

The newer membrane materials discussed in this chapter have evolved as solutions to problems with traditional built-up waterproofing membranes. Usage of built-up waterproofing membranes has declined dramatically for two major reasons. They are labor-intensive, field-fabricated products that require diligent installation. Second, their bitumen-heating kettles foul the air with hazardous hydrocarbons that are banned in many places by pollutant-abating ordinances. This chapter covers modified bitumens, single-ply elastomeric and thermoplastic sheets. LAMs; and bentonite, which is not a membrane; are covered in Chapters 11 and 12.

A. Polymer-modified bitumens

Polymer-modified bitumens can be considered an upgrade from built-up below-grade waterproofing membranes, as a single manufactured sheet (generally 60 mils thick) serves the same function as the three- to six-ply, field-fabricated traditional membranes had. Compared with built-up membranes, modified bitumens offer much greater flexibility and breaking strain, with comparable toughness and puncture resistance. For clients who want to exploit the benefits of progress without abandoning traditional waterproofing technology, modified bitumens offer a comfortable compromise; less radical than single-ply elastomeric or thermoplastic sheets. Via its many virtues, modified bitumen waterproofing enjoys skyrocketing popularity.

Like modern single-ply sheet materials, modified bitumens are a triumph of polymer chemistry, popular in both roofing and waterproofing applications. Various polymers – atatic polypropylene (APP), sequenced-butadiene styrene (SBS), styrene ethylene butylene styrene (SEBS), and polyisobutylene – are compounded with asphalt (or in rare instances, coal tar pitch) to make it more viscous, less temperature-sensitive, and more elastic. Asphalt is a viscoelastic solid at low temperatures and a viscous fluid at high temperatures, with radically different properties (e.g., breaking strain, tensile strength, thermal coefficient of expansion/contraction). Modified bitumen has superior properties, up to 150 percent breaking strain versus 2 percent for built-up membranes, plus much higher cold-temperature flexibility. Many waterproofing consultants believe membranes should have a modulus of elasticity less than 80,000 psi to maintain flexibility. Modified bitumen can resist tensile and flexural stresses that would rupture an asphalt (or coal tar pitch) built-up membrane.

Modified-bitumen waterproofing membranes are suitable for all three components: structural slabs, slabs-on-ground, and foundation walls. Their uses include all the applications previously dominated by hot-mopped, built-up waterproofing membranes. Modified bitumens combine two big labor-saving advantages in the field: (1) substitution of a single, prefabricated sheet for several felts incorporated into built-up membranes, and (2) cold application instead of built-up membranes' hot application, with its labor-intensive hot kettles and air-polluting fumes.

Modified bitumens offer special benefits on vertical surfaces, where the ease of hanging single sheets of factory-controlled thickness eliminates the problem of hot-mopping multiple plies of built-up membrane or brushing, troweling, or squeegeeing two layers of LAM (see Figure 10.1).

Modern modified bitumen membranes have evolved from a 70-mil sheet introduced roughly 45 years ago by W.R. Grace, under the proprietary name, Bituthene. Like contemporary modified bitumen products, Bituthene was rubberized asphalt, modified by reclaimed rubber and resin modifiers, laminated to a special cross-laminated HDPE film. After Grace's patents had expired, other manufacturers began marketing similar products, but

Figure 10.1 Modified bitumen sheets on vertical wall

Table 10.1 Suggested prescriptive and performance criteria for modified-bitumen waterproofing membranes

Property	*Test method (ASTM, except as noted)*	*Typical value*
Thickness (mils) (including 4 mil HDPE)		60
Flexibility, 180° bend over 25mm (1 in.) mandrel at -43°C (-45°F)	D 146 or D 1970	Unaffected
Tensile strength membrane (PSL)	D 412 Die C Modified	250–320
Elongation, ultimate failure of rubberized asphalt	D 412 Die C Modified	300
Crack Cycling at -32°C (-25°F), 100 Cycles	C 836	Unaffected
Lap adhesion at minimum application temperature	D 1876 Modified	4 to 7 lb./in.
Peel Strength 16/in.	D 903	7.5 to 9
Puncture resistance membrane (lbs.)	C 154	40
Resistance to hydrostatic head (ft.)	D 5385	150 to 230
Exposure to fungi in soil, 16 weeks	GSA-PBS 07115 16 weeks	Unaffected
Permeance (perms)	E 96 Method B	.05
Water absorption (by wt.)	D 570	0.1 to 0.25

with different asphalt modifiers. These self-adhering sheets require substrate priming to complete the chemical curing that assures a tight adhesive bond at the membrane-substrate interface. They also require pressure rolling of field seams, to assure tight adhesion at these vulnerable joints.

The most popular modified bitumen waterproofing membrane is a self-adhering, 56-mil sheet laminated to a 4-mil HDPE film. It is called self-adhering rubberized asphalt. The HDPE film provides multiple benefits:

- increased tensile strength and breaking strain
- resistance to acid soils and organic growth
- improved resilience, self-healing property, and bondability
- easier joint seaming
- improved resistance to vapor flow
- enhanced crack bridging
- improved bonding adhesion

These self-adhering sheets have a silicone release paper on the modified-bitumen side, designed to prevent sticking (see Figure 10.2). System accessories for installation include a solvent primer (some in a VOC-compliant formulation), a water-based primer, mastics, and trowel-grade sealants generally called "liquid membranes."

Like every other waterproofing material, modified bitumen has limitations:

- It is unsuitable for blind-side application.
- It does not adhere to pressure slabs when applied to a mud slab.
- It has poor ultraviolet-radiation resistance (most waterproofing sheets do), and thus should not be exposed to the atmosphere for more than a few months.
- Its application is limited to temperatures of 25°F and higher.

Several manufacturers of modified-bitumen sheets market products for waterproofing that were previously limited to roofing applications. These expanded waterproofing uses are generally limited to plazas and foundations.

You should be aware of the more rigorous requirements that may disqualify modified-bitumen roofing sheets as waterproofing membranes. Excessive water absorption is a major factor disqualifying most APP and

Figure 10.2 Self-adhesive membranes

SBS-modified bitumen roofing membranes as waterproofing membranes. To qualify as waterproofing membranes, modified bitumen sheets should have a maximum water absorption of two percent (by weight) per ASTM standards. Test data indicate that absorption rates for these smooth-surfaced, modified-bitumen roof membranes range from a highly satisfactory one percent or less up to a highly unsatisfactory 10 percent or higher. These tests are conducted under more or less ideal laboratory conditions on membrane specimens with sealed edges. The lenient laboratory conditions may be satisfactory for testing roof membranes, but not for membranes in a perpetually wet environment. Wicking of water can cause reinforced waterproofing membranes to swell and disbond. For this reason, most reinforced modified bitumen roof membranes are unsuitable for waterproofing membranes for below-grade applications.

Several cold-applied built-up waterproofing membrane systems feature materials that could be classified as modified bitumens. In one system, a polymer-modified asphalt adhesive can be cold-applied to a polymer-modified sheet reinforced with a glass mat. This system is applied shingle fashion, with one to three plies, depending on hydrostatic pressure. Plies are rolled in, hung, or rolled out on the flat. The adhesive reacts with the polymer-modified sheets and fuses them. After curing, the individual plies are indiscernible; the membrane becomes an integral, reinforced sheet rather than a multi-ply, built-up laminate. As a major advantage, this system can be applied to damp surfaces.

B. Elastomeric and thermoplastic membranes

Ethylene propylene diene rubber (EPDM) is the principal single-ply elastomeric sheet used in below-grade waterproofing. It is 60 to 120 mils thick and is installed loose-laid or fully adhered on horizontal surfaces and fully adhered on foundation walls. Butyl is no longer available in large sheets. Manufacturers no longer recommend intermittently attaching EPDM on vertical surfaces. Backfilling and settlement tended to drag the membrane down and disbond it from the intermittent attachments.

PVC sheets have the essential properties of low absorbance (two percent) and low permeance, which makes PVC and its ketone ethylene ester (KEE) or ethylene interpolymer (EIP) blend the only thermoplastic materials recommended for waterproofing. Chlorsulfonated polyethylene (CSPE, trademarked as Hypalon by DuPont), thermoplastic polyolefin (TPO), other plastic roofing sheets, and laminated PVC are unsuitable for waterproofing because of their high permeability, high water absorption, or both. However, one manufacturer markets a TPO sheet coated with a butyl alloy for use on blindside applications that it claims has superior resistance to absorption and permeance.

PVC waterproofing differs from PVC roofing in both thickness (59 to 120-mils for waterproofing versus 45 to 60 mil maximum for roof

Table 10.2 PVC prescriptive criteria (Sarnafil, Inc.)

Parameters	*ASTM test method*	*Typical physical properties*
Color	–	orange/grey
Overall thickness mm (inches), min.	D-751	3.00 (.120)
Tensile strength, psi, min.	D-638	1,600 psi
Elongation at break, min.,%	D-638	300 MD 280 CMD
Seam strength, min. % of ten. strength*	D-638	90%
Retention of properties after heat aging	D-3045	–
Tensile strength, min. % of original	D-638	95%
Elongation, min. % of original	D-638	95%
Tear resistance (lbf)	D-1004	35
Low temperature bend (–40 F)	D-2136	pass
Linear dimensional change, max. %	D-1204	0.01
Weight change after immersion in water, maximum, %	D-570	2.0
Puncture resistance	Federal test method 2065	66.74

Note: * Failure occurs through membrane rupture, not seam failure.

membranes) and also in composition. Additives have been introduced to resist algae and alkalis. UV stabilizers and high-temperature inhibitors have been removed.

Both PVC and KEE offer excellent resistance to bacteria, fungi, and soil chemicals. EPDM, however, can deteriorate in contact with petroleum-based poisoners and oils. Hydrocarbons can similarly promote deterioration of PVC.

PVC and KEE owe their high tensile strength chiefly to the reinforcing fabric. Tensile properties are also enhanced by long chain molecules, built-up from monomers (the basic molecular units). Polymerization increases monomeric molecular weights from the 30- to 150-pound range, to so-called macromolecules, 100 to 10,000 times as large.

A major chemical distinction differentiates thermosetting synthetic rubber sheets from thermoplastic polymer sheets. Thermosetting materials harden permanently when heated, like an egg. Thermoplastics soften when heated, like butter, and harden when cooled. With thermoplastic materials, thermal cycling can repeatedly change their physical characteristics; viscosity changes accompany temperatures changes.

This contrasting behavior stems from a basic difference in molecular structure. Thermosetting resins start as tiny threads. Heat promotes chemical reactions that cross-link these tiny molecular threads, creating a permanently rigid matrix. This molecularly cross-linked structure makes a thermosetting material more resistant to heat, solvents, general chemical attack, and creep (plastic elongation under sustained stress). Unlike thermoplastics, elastomers elongate *and* recover their original shape as well, thus accommodating stress concentrations that can split built-up bituminous membranes. Their breaking strains range upward from 300 percent, more than 100 times that of built-up membranes.

Thermoplastics comprise long, threadlike molecules, so intertwined at room temperature that they are hard to pull apart. When heated, they slide past one another, like liquid molecules, and the material will elongate (creep) under constant stress.

Thermosetting materials might seem superior, but thermoplastic materials offer superior, easier joint-sealing processes in the field. Thermosetting synthetic rubbers require use of self-adhering tapes or contact adhesives, which require a wait before the field-seaming process can be completed. Field-fabricated seams never attain the tensile strength of the sheet material. PVC seams, in contrast, are as strong as the base material. They are welded with hot air at temperatures above 1,000°F. PVC and KEE are available in up to 20-foot by 100-foot sheets, thus minimizing field seams.

Though generally less environmentally hazardous than hot kettles for built-up bituminous membranes, solvent-based primers and adhesives for field joint-seaming in EPDM membranes may exceed VOC limits.

Most PVC sheets are reinforced with non-woven glass fiber felts, which increases both tensile strength and puncture resistance.

Another category of waterproofing membrane consists of a thermoplastic, HDPE or TPO, coated with an adhesive. This should not be confused with thin (15 mil) HDPE or thick (60 mil) PVC sheets coated with a layer of modified bentonite. Both are intended for use as pre-applied membranes. The first type is designed to adhere to pressure slabs on grade and foundation walls both chemically and from the pressure exerted by the weight of the freshly cast concrete (see Chapters 8 and 12). The second type relies on the bentonite swelling and forming a water impervious gel between the soil and the concrete (see Figure 10.3).

The adhesive coated membrane consists of either a 30-mil or 16-mil HDPE sheet with a 16-mil factory coated adhesive protected with a release film. The seams are adhered and compressed with a roller. The TPO membrane is coated with a polyisobutylene rubber. The adhesive on the TPO sheets for use under slabs is formulated to resist damage from construction traffic.

C. Application

PVC and KEE sheets are suitable for application to structural slabs, *but not* on slabs-on-ground, and foundation walls. PVC and KEE sheets can be fully adhered or loose-laid over framed slabs.

Figure 10.3 HDPE laminated to bentonite applied to a primed wall

Installation practices for EPDM and PVC on framed structural slabs are generally the same. PVC and KEE sheets are laid out and heat welded. EPDM is laid out, relaxed, and seamed with tape. Using cover tape on seams for 60-mil EPDM is desirable; it is mandatory for thicker sheets.

Loose-laid waterproofing membranes are often compartmentalized by adhering the membrane to the slab in an approximately 10-foot grid. This practice localizes leaking. It restricts the distance between the leak source (e.g., a defective lap seam) and its horizontal distance from the leak's appearance (e.g., at a crack in the ceiling surface of a framed slab). The manufacturer's water cutoff mastic is an appropriate adhesive for compartmentalizing, which blocks lateral flow of leak-water from the compartment where it has penetrated the membrane.

Flashing for EPDM membranes consists of non-vulcanized butyl sheets, which conform to their backing and later cure to attain the elastic properties of fully cured (vulcanized) material.

PVC is flashed with the same membrane sheet material; usually reinforced, except at corners and penetrations; where conformability rather than strength is a prime requirement.

Adhesive-coated HDPE and TPO sheets when installed on mud mats can have the adhesive destroyed by foot traffic (generated by placing rebar), dirt-covered puddles, and debris blown or thrown into the excavation. This can impair their ability to aggressively bond to the concrete cast over them. A three- to four-inch thick layer of concrete cast over the sheet as soon as possible will provide sufficient protection as well as a platform for construction traffic, setting rebar chairs and installing piping.

D. Modified bitumen failure modes

Modified bitumen failures can be categorized as disbonding from the substrate, defective seams, and ruptures. Disbonding is the most prevalent (see Figure 10.4). It results in blistering, slippage, and wrinkling.

Disbonding primarily results from failure to:

- promptly cover the membrane after installation. Self-adhering modified bitumen sheets have a film of HDPE that is an effective vapor retarder. When the sheet is exposed to sunlight for relatively short periods of time the black surface temperatures soar, vaporizing volatiles in the

Figure 10.4 Slippage, resulting in total disbonding of self-adhering modified-bitumen waterproofing membrane on foundation wall, is accompanied by delamination of drainage composite

primer. (Applying the primer too thickly or installing the sheet too soon aggravates this situation.) The HDPE prevents the release of the vapors, resulting in blistering and disbonding. This condition is frequently found at brick shelves where masonry erection is often delayed for weeks or months.

- seal the top edge of vertically applied self-adhering sheets at day's end. This permits moisture to flow behind the sheet and disbond the top edge. Dust and dirt from backfilling operations inhibit adhesion. When the next sheet is applied, there is a continuous un-adhered strip.
- apply primer correctly – e.g., failure to reprime if the sheet is not installed the same day; to wait until the primer has dried, with all solvents flashed off; or to use the appropriate grade, consistent with the ambient temperature.
- fill tie rod holes, bug holes, and others until they are flush with the concrete surface.

Defective seams primarily result from failure to:

- roll the seams or exert sufficient pressure promptly after application.
- cut out fishmouths rather than trying to flatten them.
- align the sheets correctly. Aligning them to the height of the wall can cause backfill settlement to disbond them.
- properly calibrate welding machines.

An allied problem is the lack of patches over tee joints.

The most common cause of **rupture** is a failure to reinforce dynamic joints – e.g., where a structural slab bears on a foundation wall.

E. Checklist: modified bitumens

1. Beware of modified bitumen roofing sheets promoted as waterproofing membranes. Check their water absorption rates and their vulnerability to wicking of moisture through their edges.
2. Require inspection of all modified bitumen installations, with special attention to pressure rolling of membrane and field-splicing seams, to assure tight bonding adhesion.
3. Require sealing of top edges at end of day's work.
4. Use liquid waterproofing on vertical and horizontal reentrant angles to seal between sheet and substrate.

F. Checklist: PVC

1. Assure that bonding adhesive is kept off seams.
2. Check PVC seams at beginning of each day and after lunch to verify welder settings (calibration).

3. Verify that substrates are smooth and clean.
4. Walk each seam and check with a pointed tool at end of day's work.
5. Walk seams after membrane verification testing.

11 Liquid-applied membranes

A. Description, types, risks

LAMs are field-fabricated from a variety of liquid components applied to the substrate via rollers, brush, spray, trowel, or squeegee (see Figure 11.1). They may be cured via chemical reaction, exposure to moisture, solvent evaporation; or hot-applied and then allowed to cool. Two coats are often applied.

Figure 11.1 Notched squeegee applies single-component, moisture-cured tar urethane film, one of several methods – also including rolling, brushing, and spraying – for installing liquid-applied membranes (LAMs) to concrete substrates

The major distinction between LAMs is whether they are hot-applied or cold-applied. Hot-applied LAMs have historically been superior to cold-applied. They:

- cure faster (almost as soon as they cool);
- are usually reinforced (unlike many cold-applied LAMs); and
- are roughly three times as thick as cold-applied systems, thereby providing superior puncture resistance, tensile strength, and other desirable properties (see Figure 11.2).

According to Richard Boon, roofing consultant, cold-applied LAMs are dependent on a complex chemical curing process. When cleanliness, air temperature, and humidity were less than ideal, previous generations of cold-applied LAMs did not cure properly. Advances in chemical engineering have improved the process. Added reinforcements are also increasing the reliability of cold-applied LAMs.

Figure 11.2 Reinforcing fleece is shown being embedded in the first application of rubberized asphalt while it is still hot. A second coat will be applied over the reinforcing to complete the membrane. The hot membrane will be covered with a protection ply to adhere it, but the side laps will not be sealed

Both systems are susceptible to moisture in the substrate and may require special primers when the moisture content exceeds the manufacturer's specified limits. This is particularly challenging when applicators begin their work early in the morning when the concrete substrate is covered with dew.

Both systems require more rigorous field controls than alternative waterproofing systems. These requirements start with substrate preparation, which demands fastidious care and cleanliness, a challenge to construction workers' apparent religion. Weather conditions are critical during application, and the application process itself places further demands on the applicator's abilities. Thus, a reliable contractor, backed by a good quality assurance (QA) program, is essential to produce a dependable LAM system.

If you put a coating in a can, someone probably will open the can and improperly use the contents. For many of these moisture-cured products, just opening the can will trigger the chemical reaction that begins the curing process.

Despite these disadvantages, liquid-applied systems are sometimes the best choice when you must rule out prefabricated-sheet systems. Their self-flashing property and their availability in thixotropic grades suitable for vertical surfaces make liquid-applied systems suitable for projects with complex geometric contours; multiple plane transitions; and numerous pipe, duct, and other penetrations. Any one of these complicating factors can disqualify conventional, prefabricated-sheet waterproofing systems from consideration.

Determining the precise degree of complexity needed to establish the viability of a liquid-applied membrane for a specific project is a risk-balancing exercise. Many changes in a structural slab's elevation, unsymmetrical or irregular flashing contours, or penetrations through slabs and walls make the flashing details of a prefabricated-sheet system less practicable than the extra care required in preparation and application of a liquid-applied membrane.

B. Historical background

LAMs were first introduced as waterproofing systems at least as far back as the early 1960s. According to the late Philip Maslow, an expert on construction sealants and waterproofing techniques, pour-grade polysulfide sealants provided the stimulus to expand this material's use into LAMs. The polysulfide sealant formulation was modified to increase its breaking strain, as a means of preventing shrinkage cracking in a concrete substrate from propagating through the adhered LAM.[1] In the pioneering LAM formulations, liquid polysulfide polymers were blended with coal tar oils.

Several fatal flaws soon led to the disappearance of polysulfide systems as LAM waterproofing. Their sensitivity to temperature posed a major obstacle. At high temperatures, they cured too quickly; at low temperatures, too slowly. Their pot life was also too short.

Liquid or fluid applied waterproofing membranes have been marketed since the late 1970s but became popular in the 1990s when a variety of cold-applied membranes were introduced.

Polyurethanes, similarly blended with coal tar oils, are not temperature-sensitive like polysulfide polymers, but they are *moisture*-sensitive. This problem was partially resolved with the introduction of epoxy primers.

In the past decade, two new LAMs were introduced to the plaza waterproofing market: Two- and three-component polyester and two-component polymethyl methacrylate (PMMA and MMA). Their outstanding advantages are their self-flashing capability and quick drying. Because of this, some manufacturers of hot-applied rubberized asphalt and built-up modified bitumen membranes permit its use for flashing their systems. Both are resistant to most chemicals found on job sites, but the PMMAs are sensitive to substrate moisture and the polyesters to VOC limits.

C. Materials

Liquid-applied waterproofing membranes come in the following types:

- hot- or cold-applied polymer-modified asphalt
- single-component, modified polyurethane
- single-component, unmodified polyurethane
- two-component polyurethane
- two-component, modified polyurethane
- two-component, unmodified polyurethane
- two-component, unmodified latex-rubber
- two- and three-component, polyester
- two-component polymethyl methacrylate (PMMA and MMA)

This list *excludes* unreinforced asphalt cutbacks and emulsions that cure via evaporation of water or petroleum solvents. Those materials qualify only for *dampproofing*, not *waterproofing* (see Chapter 3 "Dampproofing"). The late Carl Cash claimed that any unreinforced coating was dampproofing. Epoxy and polyurea are also excluded; they have limited use as stand-alone membranes at this writing.

Compared with single-component membranes, two-component membranes offer several notable advantages:

- fewer weather restrictions on application;
- more predictable curing; and
- longer shelf life.

These advantages stem from the two-component coatings' chemical curing, which makes them less affected by site humidity. They can be applied at lower ambient temperatures than single-component coatings and will cure at lower temperatures. Most single-component coatings require exposure to air with limited absolute humidity to assure proper curing.

Single-component coatings' curing rates are less predictable than those of two-component coatings. Most of these materials rely on atmospheric humidity to provide the water that acts as a catalyst. This exposes them to construction damage avoided by two-component systems. Premature installation of protection boards on single-component coatings can retard – or at worst, prevent – complete curing. Many protection boards are impermeable. Incomplete curing impairs waterproofing integrity. Moreover, blowing dust, rain, salt-laced fog at coastal sites, and traffic can destroy uncured coatings.

Its one comparative disadvantage is that two-component coatings require thorough field mixing to assure complete blending. Although they have a longer shelf life than single-component coatings, observance of a two-component coatings' shelf-life limit is even more critical than for single-component coatings.

The tradeoffs of construction convenience versus dependable curing are also exemplified by the few liquid-applied coatings that require jacketed (double-boiler) kettles on the jobsite for hot application. Polymer-modified asphalt coatings designed for hot application require these more expensive kettles, which may also require the contractor's maintaining a fire watch after the last flame is extinguished. As a counterbalancing advantage, hot-applied coatings offer almost instantaneous curing.

In contrast, polyurethane-based cold-applied LAMs require much longer curing times; in extreme cases up to 72 hours at 75°F (and more at lower ambient temperatures). Cold-applied LAMs require at least 12-hour (overnight) curing time (per coat) at 70°F, 50% RH, (longer at lower temperature or lower RH). Two-coat, cold-applied membranes may require 48-hour curing time to assure evaporation of solvents in the underlying coat. Conversely, PMMA and polyester membranes cure very quickly, which can be less than a day. However, curing times for the membrane are in addition to the curing time required for the primer which may be as long as overnight. Pinholing can occur if curing time is inadequate.

Hot-applied membranes are much less vulnerable to sudden temperature drops, rain, or premature traffic than cold-applied LAMs are because they don't require lengthy variable curing times. This gives them a huge advantage.

Polyester waterproofing, currently produced by one manufacturer, is a fleece-reinforced polyester resin-based two-component waterproofing system generally used for vegetated roof and IRMA roofing assemblies. It can be laid under asphalt, slabs, or paving. Poured asphalt can be laid at 480°F without damaging the membrane. It is permanently elastic and bridges cracks up to two mm, making it an ideal waterproofing solution for surfaces with vehicular or pedestrian traffic. It is also vapor permeable within a few hours of application. Thereafter, the water vapor transmission (WVT) drops to 0.27 perms.

Poly methyl methacrylac (PMMA or MMA) is currently produced by at least four manufacturers, one of which is in the U.S. It can be reinforced when used over occupied space and unreinforced on balconies. Where used below grade or on plazas, reinforcing is mandatory. PMMA is rapid drying and can be walked on within a day of its application. Its outstanding advantage is that it is self-flashing.

D. LAMs versus other waterproofing systems

In comparing LAMs in general with other types of waterproofing systems, you can follow these preliminary design rules:

- Specify LAMs for projects with many penetrations.
- Specify LAMs for projects with complex contoured surfaces and plane changes.
- Specify LAMs for new construction with cast-in-place structural concrete decks (minimum 110-pcf concrete density).
- Favor urethane LAMs where resistance to chemical attack is a major design requirement.
- Use LAMs on horizontal surfaces (i.e., plazas, structural slabs, and slabs-on-ground) rather than vertical surfaces (where it is difficult to assure adequate and uniform thickness).
- Favor reinforced LAMs over unreinforced LAMs.
- Ensure that a selected LAM can meet VOC requirements where required.

LAMs are less suitable or unsuitable for certain uses. Avoid or be wary of specifying LAMs for:

- rough, irregular (i.e., non-planar or ripply) surfaces. Rough surfaces make it difficult to achieve minimum thickness and require additional primer. Unless the LAM material is thixotropic (meaning it becomes fluid when agitated and subsequently reverts to viscous state) and trowelable, two coats are normally required to assure minimum thickness.
- uses where LAMs' relatively limited crack-spanning ability could be a concern.
- remedial waterproofing projects because the substrate may be contaminated with incompatible materials or too coarse for economical application.
- application over lightweight insulating concrete, precast concrete or pre-stressed concrete units, or concrete fill over these members. Lightweight insulating concrete fills are subject to excessive shrinkage cracks, which can propagate upward into the LAM.
- on concrete decks over non-vented steel forms. Non-vented decks can blister the LAM with vapor-pressure buildup resulting from excess moisture retention. LAM substrates require a relatively dry substrate.

- where tight construction schedules may force application before the concrete cures, and before the inevitable shrinkage cracking occurs. Super-plasticized or high-early-strength concrete, and concrete with water/cement ratios of greater than 0.40 can minimize this problem. However, they cost much more.
- in confined spaces where VOC limits or fume toxicity are limiting factors.
- filling or leveling surface irregularities.
- under a continuous head of water. A compression of 95 percent maximum dry density per ASTM D1557 or greater acting on an unsupported membrane can cause it to deform, extrude, or even rupture.
- over high humidity spaces such as kitchens, pools, laundries, etc.
- on, or even in close proximity to, occupied buildings. Moisture-cured, coal-tar-pitch modified urethane and polyester LAMs can emit odors in excess of VOC limitations and might be toxic to some people during the curing stage. Since LAMs are especially popular on waterproofed plazas; where air intakes, ventilators, and adjacent windows may admit fumes into the building, you must be especially alert to toxic emissions in such applications. Check with the manufacturer to avoid these curing-stage emissions.

Don't specify hot-applied LAMs on walls, unless they are parged. (Vertical surfaces make control of uniform thickness difficult.)

Don't specify cold-applied LAMs on unreinforced masonry walls. The parging cracks. Don't specify cold-applied LAMs when the ambient temperature may fall below 45° F during application, unless the product is designed specifically for low-temperature curing. Hot-applied LAMs may be acceptable. Application to frozen substrates is never a good idea.

E. General requirements

After long experience, the industry has established prescriptive criteria for minimum solids content (by weight) and dry-film thickness (DFT) for cold-applied LAMs.

Minimum solids content is 80 percent. Solids mean the residual material remaining permanently in place after initial evaporation of solvents. This standard is included in specifications to reduce the probability of pinholing and to help assure a minimum thickness of coating after the carrier has evaporated. Pinholing is a major defect that can impair the waterproofing integrity of liquid-applied coatings, as can thin applications. As discussed in Section H, pinholing and blister are the most prevalent defects in liquid-applied waterproofing membranes.

Minimum coating thickness (DFT) is 60 mil. This standard averts several hazards associated with liquid-applied coatings. Substrate cracking, propagated vertically from a concrete slab or horizontally through a concrete or masonry wall, can produce stress concentrations in thin coatings where irregular substrate reduces cross-sectional thickness (see Figure 11.3). Thicker membranes are also less susceptible to wrinkling from absorbed moisture, which expands the membrane and disbonds it from its substrate. Thicker membranes also reduce the incidence of pinholing, by reducing the probability of voids extending through the entire membrane cross section. Note that this standard is double the 30 mil thickness established for LAMs on sprayed polyurethane foam roofs.

When considering a liquid-applied waterproofing coating, note that uniform coating thickness is easier to achieve on horizontal surfaces than on vertical surfaces. Gravity is a friendly force on slabs, an enemy on walls. This disparity is accentuated for hot-applied coatings, which are less thixotropic than cold-applied coatings.

General criteria for cold-applied LAMs are listed in ASTM C836 "High Solids Content Cold Liquid-Applied Elastomeric Waterproofing Membrane for Use with Separate Wearing Course." The criteria include:

- hardness
- weight loss (20 percent = 80 percent solids)
- low-temperature flexibility
- low-temperature crack bridging
- adhesion in peel after water immersion
- extensibility after heat aging
- stability

With few exceptions, all LAMs are sensitive to moisture in the substrate and require careful testing prior to priming to ensure that the surfaces are dry.

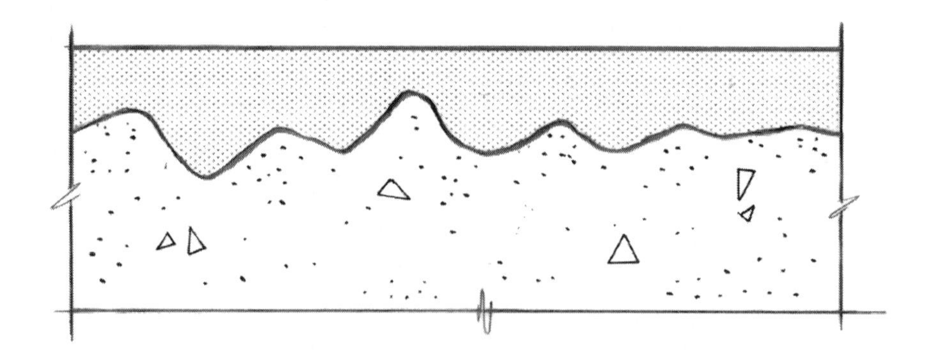

Figure 11.3 60-mil. minimum thickness is required for LAMs to avoid excessively thin cross sections over rough, irregular substrates. Stress concentrations can split the membrane at thinned cross sections, thereby destroying the membrane's waterproofing integrity

F. Testing for dryness

Manufacturers of LAMs always specify that the substrate is dry. However, many do not specify "how dry is dry," how to measure dryness and what the acceptable limits are. ASTM E1907 "Standard Test Methods for Determining Moisture Related Suitability of Concrete Floors to Receive Vapor Retardant Finishes" describes six tests to measure the readiness of concrete slabs to receive vapor retardant finishes. They are polyethylene sheet test, mat test, electrical resistance test, RMA test, primer or adhesive strip test, and humidifier test. Some of these tests take more than 48 hours to complete and are not feasible for slabs that are exposed to the elements. The most common test is the polyethylene sheet test (see ASTM D4263 "Standard Test Method for Indicating Moisture in Concrete by the Plastic Sheet Method") that can be completed in 18 to 24 hours. However, the results are not reliable as they simply indicate whether moisture was visible under a plastic sheet taped to the concrete. In my opinion, a more reliable test is to coat a small area with the specified coating system and try to pull it up after 12 hours (see Figure 11.4).

Figure 11.4 A sample patch of hot rubberized asphalt membrane is installed on a plaza deck to evaluate its adhesion and the dryness of the substrate. Coupons will be extracted and inspected for interplay adhesion, adhesion to the deck, and the presence of interplay blisters

G. Application

Most LAMs are applied in a single coat to a roughly 1/16-inch-thickness, either by roller, trowel, or squeegee. Most coatings can be sprayed. Coatings with 100 percent solids (i.e., those containing no evaporative solvents) may require trowel, roller, or squeegee application. Many moisture-cured products are applied in a liquid state. As warned in Section A, liquid-membrane application requires much greater care and considerably greater skill than built-up, modified bitumen, or prefabricated, single-ply sheet installation.

Reinforced membranes are superior. A two-coat application reduces the potential for pinholes and thin spots.

Avoid low-solids, two-coat, cold-applied single-component reinforced membranes. A low-solids LAM on reinforcing may cause pinholing because the cold-applied liquid tends to flow in between the strands of reinforcing mesh, and the reinforcing may inhibit moisture curing. Hot application avoids this problem. Its rapid change from liquid to solid promotes a smooth, plane surface over the reinforcing mesh.

Substrate irregularities cause variations in coating-film thickness. Frequent checks with wet-mil gauges are thus required during installation operations to assure maintenance of at least 60-mil thickness. Coupons should be cut from each square (100 square feet) of cured membrane and measured for thickness.

Where membranes are found to be below gauge, either in film thickness or where cratering is extensive, additional coats are required to provide the minimum thickness throughout. Most LAM materials readily bond to themselves when cured, providing the prior coat is clean and dry. Some require a primer.

Concrete substrates for liquid-applied membranes must be rigorously prepared: fully cured and free of adhesion-inhibiting curing agents, sharp fins (see Figure G.1), and other irregularities.

Most manufacturers require a steel-trowel finish or wood troweling plus a lightly broomed hair broom finish. When the surface is cured to the manufacturer's satisfaction, it is swept clean or preferably vacuumed. Cracks and joints should be air-blasted.

Virtually all LAMs require a primer. Primers must be water-based where solvent-based primers are not VOC-compliant. Epoxy primers are frequently used where moisture is present in the substrate.

Cracks and joints must be stripped with woven glass strips adhered with liquid-membrane material, or caulked and then stripped. Flashing of liquid-applied systems with a thixotropic version of the membrane liquid is a simple operation. For LAM systems with sheet flashing – neoprene rubber or polymer-modified bitumen – flashing becomes a more complicated problem. EPDM flashing may prove unsuitable for use with bitumen-based LAMs because it extracts less of the oil in the bitumen.

H. Failure modes

Common defects in liquid-applied membranes are:

- pinholing
- cratering
- wrinkling
- blistering
- splitting
- delamination

Pinholing, the most common defect, has several possible causes. A wet, dusty substrate can produce pinholes. Urethane is especially sensitive to moisture, which can set off a carbon-dioxide foaming reaction. Concrete curing compounds on the substrate – sodium silicate or animal fats – can also produce pinholes. In addition to chemical reactions and inadequate substrate preparation, faulty mixing of a two-component coating can entrap air bubbles within the membrane film. These entrapped air bubbles can either reduce film thickness or extend through the entire membrane cross section.

Solvents within a LAM can also cause pinholing. In a poor LAM formulation, the liquid can form into a premature gel state at the surface. Solvent escaping from the bottom of the membrane can then bubble up from the membrane cross section, creating pinholes in the surface gel stratum.

Pinholes are difficult to repair with high-viscosity coatings. The remedial coating bridges the pinhole; the coating dries from top and bottom to form a concave lens cross-sectional shape over the pinhole; and the lens-shaped cross section can break at its thinnest point, thereby recreating the pinhole.

Cratering forms in a manner similar to pinholing. Entrapped air bubbles exposed to solar heat expand and burst, leaving craters in the membrane's top surface. When protection boards are prematurely placed on undercured, single-component LAMs, entrapped air bubbles form depressions in the film, with consequent drastically reduced thickness.

Both pinholing and cratering can be limited, or even eliminated, by backrolling or using a spike roller on lower-viscosity liquids. Weather-sensitive scheduling can also prevent solar expansion of entrapped air bubbles by assuring falling ambient temperatures and solar-radiation intensity, both of which usually decline in the later afternoon hours (in some parts of the country). Afternoon scheduling also averts the problem of morning condensation on the slab.

Wrinkling is usually caused by continuing degeneration of a membrane already damaged by pinholing. Under low hydrostatic pressure, surface tension will normally be high enough to prevent water passage through a pinholed membrane. Under a hydrostatic head, however, water will readily leak through pinholes. Once it penetrates the membrane, water will migrate along the substrate at the interface plane between substrate and membrane.

Absorbing water from both sides, under-gage membranes swell and disbond from the substrate. The lengthening membrane buckles under compressive stress induced by adhesive restraint at the substrate. This swelling-caused disbonding process produces a "wrinkled brain" appearance (see Figure 11.5).

Wrinkling can also be caused by a solvent that expands the liquid membrane's volume. You can test adhesion of a first coat with a solvent. If it wrinkles, it obviously has not adhered. The crests of the wrinkle waves indicate a disbonded membrane directly underneath. If it doesn't wrinkle, it is adhered. This adhesion test provides another advantage for two-coat LAMs: application of a second coat tests the adhesion of the first coat.

Blistering occurs during, or soon after, application, and usually results in cratering where the blister bursts (see Figure 11.6). Blisters are formed when moisture migrates up through the concrete after the membrane is partially or fully cured. The cause is premature LAM application. LAM application should be scheduled after the concrete has completed most of its curing; at least 28 days after concrete casting, or when it is tested per ASTM E1907, with the approval of manufacturer and applicator.

Blistering is a manageable problem if it occurs during application and if it produces craters. If it occurs after installation of the protection board and overburden, it is a severe problem.

Splitting in LAMs can occur over unrepaired cracks in the concrete substrate. An unrepaired concrete crack concentrates membrane strain (and consequent stress) as the crack opens or closes in response to temperature, moisture change, or structural deflection. Splits can also occur at substrate cracks that develop after installation of the LAM.

Another common location for splits is at joints between sheet flashing and the LAM membrane. Stress concentrates at these joints because the

Figure 11.5 The "wrinkled brain" pattern of the failed hot-applied, modified-bitumen LAM (left) indicates swelling caused by moisture absorbed from concrete below. The rectangular ridge pattern was formed by protection boards. Surface moisture was evident when membrane was test cut (right)

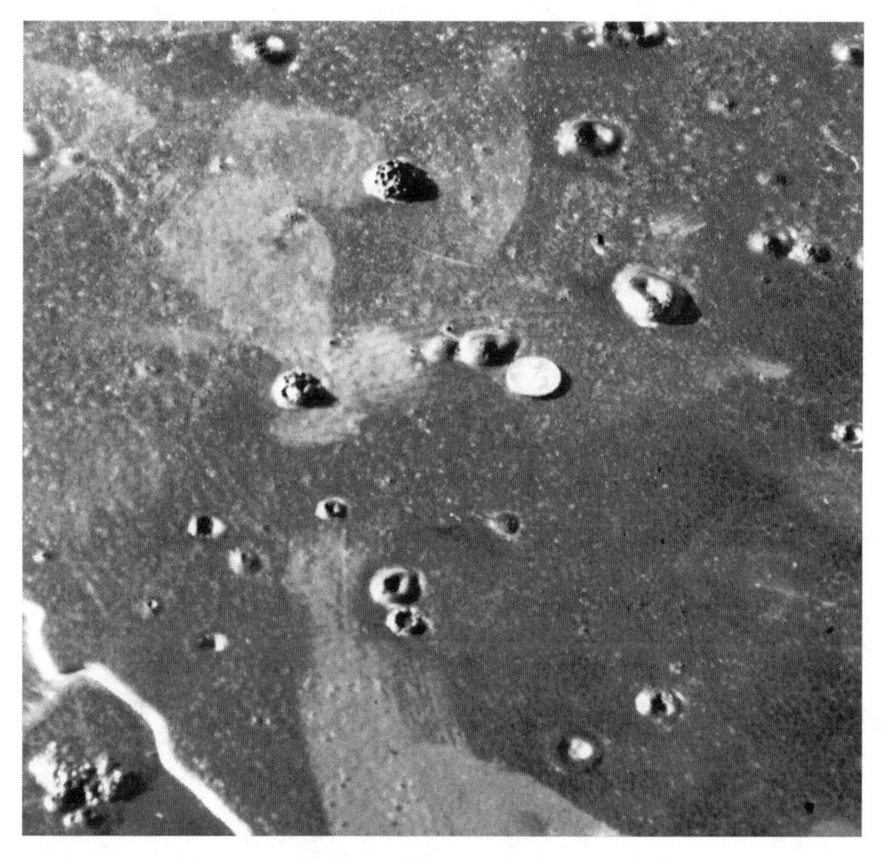

Figure 11.6 Blisters and craters indicate poor surface preparation or air entrainment during application. LAM should be applied on dry, dust-free surfaces

flashing is less elastic than the LAM and the membrane is often thinner (and consequently weaker) at these locations.

Delamination is disbonding of the LAM material from the surface to which it was applied, or between layers of the material. This is generally a result of dirty substrates or other surface contamination.

I. Checklist

1. **General**
   a. Follow these steps in conjunction with manufacturer's recommendations.
   b. Require certification of applicator as a qualified liquid-applied contractor.

   c. Consider including on-site inspection of application in your contract.
2. **Design**
   a. Consider liquid-applied membranes only if other membrane types are impracticable.
   b. Check with LAM manufacturers about possible emission of toxic odors in a LAM's curing stage, and avoid such products' use on occupied buildings.
3. **Field**
   a. Require pre-application approval of substrate preparation and penetration details by a manufacturer's representative and inspector.
   b. Establish a dependable QA program, preferably with on-site inspection throughout the application process.
   c. Require testing of wet-film thickness during application. Require cutting of dry-film coupons during application to test thickness Have one cut for every 100 square feet of LAM.
   d. Assure that squeegee notches are clean at the beginning of each day and after the lunch break.

Note

1 Philip Maslow, *Chemical Materials for Construction*, Structure Publishing Co., Farmington, MI, 1974, p. 464.

12 Bentonite waterproofing

Bentonite is a traditional waterproofing material whose use goes back almost a century. It has been transmuted by modern polymer chemistry into a wide spectrum of composites, which has increased its popularity and reliability. Bentonite has become a versatile waterproofing material, available in the following forms:

- interlocked geotextile sheets filled with bentonite granules high-density polyethylene (HDPE) sheets with adhered bentonite compound
- PVC or KEE sheets with adhered polymer modified bentonite (see Figure 12.1)
- remedial injection grout

Figure 12.1 KEE sheets with adhered polymer bentonite being installed blindside on a foundation wall. The KEE seams are being welded

According to Stacy Byrd of CETCO, bentonite is especially good for otherwise intractable problems – notably, blindside waterproofing in deep excavations.

But beware, not all bentonite products are comparable. Grades of bentonite, manufacturing methods, and composition of composite products vary among manufacturers.

A. Historical background

Contemporary bentonite products evolved from the basic material first used for waterproofing in the mid-1920s. Used only in granular form in those days, bentonite use was limited to sealing pond liners and compacted-earth dams (by mixing with surface soils) until the late 1950s when it was introduced to the building waterproofing market. By the mid-1960s, it was available in biodegradable kraft board panels, semi-pre-hydrated trowel-grade mastic, and sprayed formulations; the precursors of today's versatile variety of sheet products.

The replacement of traditional forms of bentonite follows the pattern set by metallic oxides' replacement by crystalline coatings and built-up membranes' replacement by self-adhering modified-bitumen sheets. Sprayed and trowel-grade membrane bentonite products have virtually disappeared from the market, a consequence of the difficulty in applying uniform thickness and a tendency to pre-hydrate. Sprayed mixtures of bentonite and asphalt also proved ineffective because the asphalt would coat over and affect the swelling performance of the bentonite.

Bentonite-filled cardboard panels were used for bentonite waterproofing from the 1960s into this century. They have been replaced by bentonite products marketed in larger composite sheet rolls, containing bentonite granules either (a) encapsulated between two polypropylene geotextiles needle-punched together, or (b) laminated to geomembranes or HDPE on one side. These new sheet composites are used for under-slab and blindside waterproofing. Needle-punched geotextile composites have shown properties that can inhibit harmful pre-hydration swelling of the bentonite from groundwater or precipitation prior to concrete placement.

Needle-punched, interlocked geotextile composites are especially appropriate for under-slab and blindside applications. They contain approximately 1 psf (4.8 kg/m^2) of bentonite granules encapsulated between support fabrics of woven and non-woven polypropylene geotextiles. In one product, the bentonite is adhered in place between the two fabrics. In another proprietary product, the interlocked, needle-punched geotextiles maintain the bentonite in place. Various geotextile composite products with geomembrane liner layers integrally bonded to one side of the membranes are also available.

Bentonite membranes that incorporate a geomembrane have lower moisture vapor permeance than membranes without the polymeric liner. One manufacturer produces a membrane with a 60-mil (1.5mm) thick HDPE geomembrane liner that can be hot-air welded (fused) together for optimal moisture vapor resistance.

Polypropylene support fabrics are porous, non-biodegradable, and non-toxic; featuring a weave that mechanically bonds them to concrete cast against them. These fabrics, when needle-punched across the bentonite layer, reportedly limit pre-hydration of the bentonite from rain or groundwater because the geotextiles restrict the bentonite from fully swelling prior to concrete or backfill placement. The geotextiles also provide sufficient shielding against construction damage to justify the elimination of protection courses, according to the manufacturers. Be wary of those claims; some types of backfill might damage membranes.

Sheets are 48 inches (1.2 m) wide, in lengths varying from 15 feet (4.6m) to 25 feet (7.6m). Membrane dry thickness is generally 1/4-inch (6mm), but varies depending on bentonite content, mass per unit area of the geotextiles, and if the composite contains a geomembrane liner.

These large prefabricated bentonite composite sheets cover up to 60 to 100 square feet. Their use greatly improves construction productivity, which reduces labor costs.

B. Physical properties

Bentonite is comprised primarily of sodium montmorillonite, an inert mineral made of microscopic clay platelets with a laminated crystalline structure. The platelets are broad, flat (typically 1 nm thick) and colloidal (i.e. water on their surfaces becomes a significant part of the structure). The structure can absorb water molecules between the platelets, causing the bentonite to expand. Upon hydration, sodium bentonite can swell 10 to 15 times its original dry volume in an unconfined condition. The absorbed water is held in place by electrical charges on the surface of the platelets; therefore, once the surface area of the platelet has been saturated, any further water is repelled. This repulsion of additional water provides bentonite with its low permeability. In a confined condition, such as below-grade, its swelling capacity is restricted and it transforms into a dense, plastic, low permeable material; which makes it a good water barrier.

To perform at peak waterproofing efficiency, bentonite must be compressed under continuous pressure to three or more times its base (i.e., unhydrated) density, a fraction of its unrestrained swollen volume. This requires a uniform compressive stress at least 40 psf. The demanding requirements for bentonite's hydration explain the frequent failure of traditional forms of bentonite to provide dependable waterproofing.

Bentonite can be hydrated and dried an infinite number of times without losing its swelling capacity. It can be frozen and thawed repeatedly without losing its ability to swell. Once hydrated below-grade, it can be expected to retain the absorbed moisture. Thus, bentonite can function in both hydrostatic and non-hydrostatic conditions below-grade, as long as it is restrained by compression. If it is exposed above grade to air and sunlight, the sodium bentonite would dry out. However, in exceptionally dry soils, the time it takes to rehydrate may decrease its absorption rate and nullify its ability to provide a watertight membrane.

Upon hydration, cohesion and adhesion properties are increased and bentonite will bond to many different materials, including concrete, stone, steel, and masonry. In a fully hydrated state, bentonite has low vapor permeance. When only partially hydrated, however, bentonite gains in vapor permeance. This decreases its efficacy as a vapor retarder, reducing its use for humidity-sensitive occupancies. Thereby, manufacturers generally add geomembrane liners to bentonite membranes to improve water vapor resistance.

Bentonite's need for hydration explains the frequent failure of sprayed mixtures of bentonite and asphalt. Asphalt inhibited the hydration process by reducing the surface contact area between bentonite particles and water molecules, thus, the bentonite could not hydrate. For this reason, this system has been withdrawn from the market.

Geotextile sheet products with bentonite contained in the composites can be used in applications with water tables that rise and fall without soil erosion but are unsuitable where underground streams flow and cause erosive migration of the soil.

The need to maintain minimum confinement pressure is a prime preliminary design criterion for bentonite waterproofing systems. Hydrostatic pressure can be reduced or lost under several conditions:

- backfill settlement can compact the soil, thereby making it less pervious to the intrusion of groundwater from the surrounding area;
- deterioration of blindside soil retention system components;
- passage of time, when the soil is already more or less impervious; and
- when sheets are rehydrated, the hydrating water can leak into surrounding soil unless this soil is water-saturated.

For horizontal components waterproofed with bentonite, sufficient concrete weight is recommended to assure adequate compressive stress. For reinforced slabs-on-ground, where waterproofing is located below the slab, a minimum slab thickness of six inches is recommended when installed over a mud slab; a minimum four-inch thickness on compacted ground. For hydrostatic conditions, the slab-on-ground must have sufficient concrete thickness and reinforcement to counteract anticipated hydrostatic pressures.

Soil character can impose limits on bentonite. Highly salinized or contaminated soils can inhibit the swelling of bentonite clay in the presence of liquid moisture. This reduction in swelling jeopardizes bentonite's unique waterproofing property. To combat these conditions, most manufacturers add polymers to the bentonite to improve its resistance to salts and contaminants. These contaminant-resistant treated products should be specified for projects in coastal areas and other locations with brackish water or soil contaminants, such as hydrocarbons. You should usually specify and require analysis of the soil's salinity and the water's chemistry, and then verify its acceptability with the bentonite manufacturer prior to material installation.

Bentonite panels faced with biodegradable cardboard were withdrawn from the market partly because of the undependable presence of sufficient bacteria in the soil to degrade the cardboard and expose the bentonite to hydration.

C. Bentonite installation

Bentonite waterproofing membranes are suitable for the following below-grade construction applications:

- slabs-on-ground;
- backfilled foundation walls;
- blindside shoring wall construction; and
- earth-covered structures.

Bentonite is not generally suitable for vegetated green roofs, but specific bentonite product composites are used to waterproof split-slab concrete decks. Bentonite membrane sheets produced by different manufacturers can vary. You should verify that the manufacturer approves a specific membrane product for each type of construction and application to be waterproofed.

Use of composite sheets has drastically simplified bentonite application techniques. Spraying and troweling required skilled applicators to assure uniform thickness, which was essential to the bentonite's waterproofing integrity. In contrast, sheets for waterproofed slabs-on-ground are simply laid on the substrate and lapped and stapled together (see Figure 12.2). Sheets are mechanically fastened to foundation and shoring walls. On vertical surfaces, sheets are oriented horizontally or vertically. Horizontal sheet orientation is preferred for the membrane course transitioning from under slab to the foundation wall because it will reduce sheet overlaps at this critical juncture. Hydration must be delayed until after backfilling or casting of slabs-on-ground. With backfilled wall applications, tapes can seal joints in sheets with geomembrane face liners. As a guideline, tape all exposed joints at the end of each day's work to protect the bentonite from precipitation.

1. Under pressure slabs

Bentonite sheets require a solid substrate. For slabs-on-ground, mud slabs are desirable, but not essential, if the subgrade consists of an aggregate or soil base that is uniform and compacted. For thick mat slabs, a mud slab is

Figure 12.2 Interlocked geotextile sheets filled with bentonite are rolled out on a compacted gravel substrate and the seams are lapped and stapled. A mud mat is preferred because adequately compacting gravel at rising walls and penetrations is difficult

recommended so that heavy rebar loads can be applied to the membrane and then transferred onto the smooth, consolidated mud slab surface. This is preferable to point-loading the membrane on an uneven substrate. For hydrostatic conditions, bentonite systems are generally installed under footings and grade beams, thereby wrapping the entire structure. This holistic coverage leaves no opening for water under hydrostatic pressure to penetrate.

Bentonite geotextile composite sheets form a mechanical bond to the underside of the slab and are less prone to physical damage than sheets with the bentonite layer exposed and not contained in a composite. Additionally, the geotextile composite sheets are more resistant to pre-hydration damage prior to concrete slab placement; because the bentonite is contained between two interlocked geotextiles. This is especially important for slab construction at elevator pits, and at slab perimeters. The edge of the sheet should extend out a minimum of 12 inches (300mm) past slab construction joints so it can be lapped by the succeeding slab section. It needs to be protected from damage and contamination caused by concrete overspray or splatter.

2. *Wall installation*

Prior to application, the concrete surface must be prepared, with the removal of fins and other projections, to provide a smooth surface to receive the application of bentonite. Honeycombed spots, rock pockets, indentations, tie-rod holes, form kickouts, and gaps at corners must be filled with cementitious grout to make them flush with the concrete surface. Bentonite mastic or bentonite powder contained in a water-soluble tube is applied around penetrations and as a fillet at the bottom of the foundation wall to footing transition; as well as a cant at inside corners up the wall (see Figure 12.3).

Use manufacturer's recommended trowelable grade materials for sealing gaps and joints in soldier piles, lagging, steel sheet piling, earth-formed shotcrete retaining walls, and auger-drilled caisson walls.

Irregular surfaces such as shotcrete and caisson walls should be grouted to provide a smooth, planar substrate for the bentonite waterproofing. A ½-inch-thick plywood facing ("boarding out") is optional for shoring metal sheet piling to provide a smooth, more uniform substrate. If plywood facing is used, the void created between the plywood and the sheet piling must be completely filled with concrete or compacted soils; the void cannot remain empty.

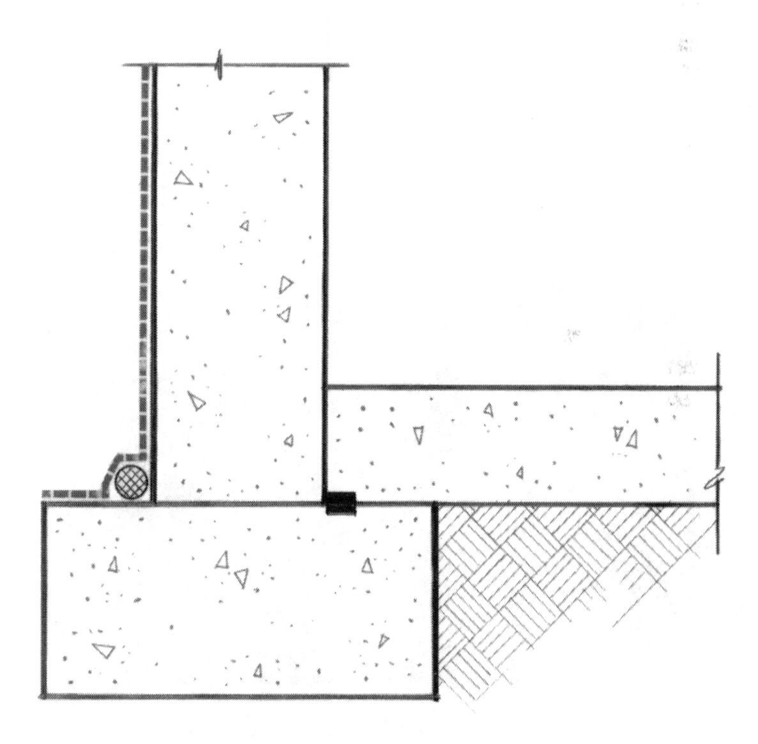

Figure 12.3 In non-hydrostatic conditions the vulnerable joint between wall footing and foundation wall should be reinforced with a bentonite-filled tube. Note: Waterstops are critical components with bentonite waterproofing

Bentonite systems pose a termination issue at grade. To maintain the required confinement pressure, they should be terminated at least six inches (150mm) below grade, with an adhered membrane flashing strip – e.g., modified bitumen or manufacturer-provided transition flashing – installed above it. The adhered flashing should (a) extend continuously from eight inches (200mm) above finished grade, and (b) overlap the top edge of the bentonite membrane by at least four inches (100mm) (see Figure 12.4). The best methods of anchoring bentonite sheets where they terminate near the top of the foundation walls are:

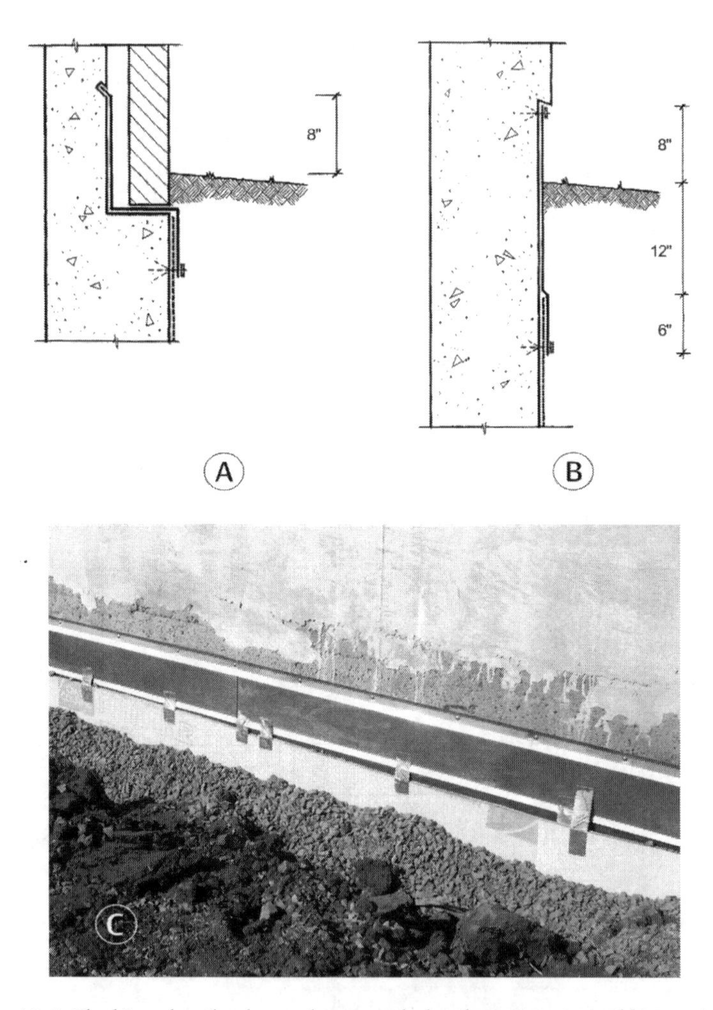

Figure 12.4 Flashing details above show grade-level termination of bentonite panels with metal flashing carried at least eight inches above grade (A) and into a grade-level reglet (B). The left detail (A) is superior. Note, however, that this detail requires level grade around the entire building

1. Mechanically fastened termination bars with bentonite mastic, or
2. Sealant tooled along the top edge of the bentonite membrane.

As applicable, install a metal cap flashing over the adhered flashing to protect it from physical damage and prolonged direct UV exposure.

Backfill must promptly follow bentonite placement, with separate backfill lifts required after installation of each sheet course. Backfill requires compaction to at least 85 percent proctor density. If backfill is not immediately placed against the bentonite sheets, temporary protection from pre-hydration against rain or snow may be required. Use plastic film, insulation boards such as extruded polystyrene, or protection boards. Damaged waterproofing materials must be replaced prior to backfill operations.

Except for geomembrane composite sheets discussed in Chapters 10 and 13, bentonite should not be used in split-slab plaza decks due to its inability to provide sufficient ballast weight and the flow of water in drainage composites.

D. Blindside bentonite

Both bentonite geotextile and geomembrane composite systems are used extensively for blindside waterproofing, notably the following zero-lot line foundation construction applications:

- concrete foundations cast against soldier piles and lagging
- steel (or wood) sheet piling
- concrete caisson retaining walls
- slurry (or shotcreted rock)
- diaphragm walls retaining earth

You must carefully detail these blockouts, leaving the waterproofing contractor no opportunity to substitute a cheaper, possibly defective, detail.

With blindside construction, waterproofing is applied directly onto the shoring wall before the building's foundation wall is formed and cast. As their major advantage, bentonite sheets can be mechanically fastened directly to shoring retention walls before the reinforced concrete foundation wall is placed. This puts the waterproofing on the positive-side of the foundation wall. Concrete cast against a bentonite-geotextile sheet forms a mechanical bond that locks the sheet to the exterior of the foundation wall. General practice with cast-in-place concrete construction is to overlap the adjoining sheet edges by at least four inches (100mm).

Cast in place concrete foundation walls exert sufficient lateral pressure to form a mechanical bond with a geotextile face membrane. Shotcrete does not. Some manufacturers have special systems for use with shotcrete.

Unlike most positive-side membranes, they seldom require "boarding out" with plywood to provide the plane, smooth substrates required for thin modified bitumens or built-up membranes. They do, however, require blockouts in walls for tieback plates that are going to be removed after the wall is cast.

Auger-cast caisson shoring walls should be grouted between piles to provide a planar surface for the bentonite sheets. Property line excavations descending into rock formations generally require the irregular rock to be smoothed with shotcrete or cementitious grout.

At the wall's base, the first sheet course should be horizontally oriented to reduce overlaps at the critical wall/footing transition. Divert jobsite water, rain, and snow melt away from the excavation until after the concrete slab and wall are cast. This warning stems from the need to delay the bentonite's hydration, especially with geomembrane types that do not encapsulate the bentonite. Removing or diverting water away from the site is a basic waterproofing principle.

E. Assets and liabilities

Bentonite waterproofing systems offer the following advantages:

- fast and relatively easy installation; can be installed onto green concrete right after the forms are removed
- suitability for blindside construction applications
- no VOC restrictions
- safe application at extreme temperatures
- relatively easy leak detection
- can bridge cracks up to 1/8-inch (3mm)
- adaptability to complex geometric shapes
- low initial cost compared with many competing materials

Fast and relatively easy installation applies to all bentonite products. Bentonite applications don't require the precise tolerances of membrane applications, where lap seams may require fastidious attention to dimensions and technique. Heat-welded geomembrane bentonite composites are an exception. Compared with other waterproofing systems, bentonite requires little substrate preparation before application.

Bentonite sheet systems are **suitable for blindside waterproofing**, notably the following zero-lot-line foundation applications:

- soldier pile and timber lagging;
- metal sheet piling; and
- auger-cast caisson shoring walls.

With blindside, the bentonite is installed before the foundation wall is poured so that the membrane is placed on the positive-side of the wall to block water.

The lack of VOC restrictions makes bentonite an option when other materials are prohibited.

Safe application at extreme temperatures, from -40°F to 130°F, generally suits bentonite for installation whenever workers can tolerate the cold or heat. In contrast, liquid-applied membranes are more severely limited by both temperature and humidity for proper curing *and* require greater care and skill by the installers.

Leak detection is relatively easy with bentonite systems because leaks normally occur near their source.

Bentonite's **ability to bridge cracks** up to 1/8-inch wide stems from its propensity to expand many times its original dry volume. This ability is, however, limited to cracks that do not allow the bentonite particles to move all the way through the crack and into the interior. Bentonite-based mastics can provide extra protection against concrete cracking by being applied to construction joints, control joints, wall-foundation joints, and other locations where cracking can be anticipated.

Adaptability to complex geometric surfaces derives from the flexibility of bentonite composite sheets. They can accommodate some geometric irregularities in the substrate. This advantage, however, is tempered by the thickness and rigidity of any geomembrane layer of the membrane. For example, membranes with a thin, flexible geotextile liner are more flexible than sheets with a 20- to 60-mil thick geomembrane.

Although **initial cost** should be your last consideration in choosing waterproofing materials, that does not mean you should always choose the most expensive method. Bentonite's relatively low initial cost is just another advantage when it is otherwise the most appropriate material.

Bentonites liabilities are:

- Its need for constant, relatively high confinement pressure to maintain the material's optimal waterproofing integrity (see Section B for details).
- Low resistance to vapor migration, unless used in a composite geomembrane such as PVC or HDPE.

Bentonite is designed to work in a confined condition, such as below-grade, where its swelling capacity is restricted and it transforms into a dense, plastic, low permeable material. It should not be used in a condition where it will not have sufficient confinement pressure from backfill or concrete topping.

Bentonite generally is less effective at retarding vapor than thermoplastic waterproofing materials are. You should specify bentonite sheets that incorporate a low permeable geomembrane component for humidity-sensitive occupancies. Seams must be carefully sealed with vapor impermeable tape to provide continuity of the vapor retarder.

When specifying bentonite, note that the wide variety of bentonite products demands close attention to the unique limitations of individual products. As with most waterproofing materials, verify that a specific bentonite product is recommended by the manufacturer for use with the relevant building material and construction. For example, designers sometimes specify the wrong type of bentonite product for masonry foundation walls, despite manufacturer's warnings limiting the product to cast-in-place concrete foundation wall applications. Additionally, bentonite is not recommended to waterproof structural wood building components below-grade. These material design oversights usually result in dismal consequences.

F. Bentonite failure modes

Though less common than roofing failures, waterproofing failures, including bentonite's, span a wider variety of locations. Roofing seldom fails in the membrane field; its major failure locations are at penetrations and terminations. Waterproofing is similarly prone to failure at these locations, but it is also vulnerable at concrete construction joints, dynamic substrate cracks, and plane changes at interior and exterior corners.

Perhaps the most vulnerable, and certainly the most difficult, location to waterproof properly is the joint at the intersection of a footing and a foundation wall. It is extremely difficult to keep the top of the footing free of dirt or mud. It is also difficult to work in the narrow trench at the bottom of an excavation well below grade, with moisture-sensitive materials lowered from the top of the excavation. Working in this cramped space on hands and knees, the applicator must place bentonite-filled tubes, with abutting ends, carefully aligned along these joints as a caulking. These challenging conditions demand that work is reviewed by the construction superintendent and the client's inspector.

Inadequate construction practice is only one source of bentonite failure. Some bentonite failures are attributable to ignoring warnings in this chapter. At an Illinois project, the fluctuating water-table elevation adjacent to Lake Michigan caused erosion of bentonite from the rise-fall cycling of groundwater. A building on a steep slope in Ohio experienced similar erosion of bentonite waterproofing. Water flowing down the slope deposited bentonite as a gray coating on the ground below. These failures may not have occurred if the designers had observed the most elementary rule in specifying bentonite: Limit its use to locations where hydrostatic pressure is stable, with no running groundwater.

Bentonite sheets (and most other waterproofing membranes) are unsuitable for use as the primary waterproofing material to seal expansion joints because joint movement induces huge strains in materials adhered to them. Instead of trying to seal expansion joints with bentonite, you should specify a properly engineered, flexible expansion joint product correctly sized to fit the joint. The expansion joint product must be anchored on both sides of the joint to prevent its separation from the concrete when the joint moves, and to form a separate watertight seal along the joint. The engineered expansion joint must also resist hydrostatic pressure normal to the expansion-joint plane. See also Appendix E: Expansion joints.

Detailing expansion joints can be difficult. Typically neoprene sheets or prefabricated metal and neoprene assembly are used to seal expansion joints. Both types require anchorage on both sides of the joint; they cannot just be stripped into the flanking bentonite. After the expansion joint assembly has been properly installed, the bentonite system should be installed overlapping the side flanges to form a continuous water barrier on the outside of the structure.

If you specify bentonite for expansion-joints, be aware that manufacturers will not warrant it for that purpose. Bentonite producers routinely warn that expansion-joint design is the "responsibility of others."

G. Checklist

1. Remember the need for continuous confinement pressure to maintain a compressive stress on a bentonite system.
2. On projects with humidity-sensitive underground occupancies, investigate the vapor-permeance properties of the bentonite system with the manufacturer.
3. Specify the bentonite system terminations a minimum of six inches (150mm) below grade with an adhered flashing strip overlapping its top edge and extending continuously for eight inches (200mm) above grade. Also, specify that tubes or bentonite mastic be installed at footing-to-foundation joints as detailing before bentonite sheet course.
4. Protect installed material from precipitation and physical damage prior to concrete or backfill placement.
5. Protect blindside installation of the product from shotcrete overspray.
6. Design a minimum four-inch-thick slab-on-ground, (minimum six–inch-thick slab over a mud slab) and verify that slab weight is sufficient to resist any anticipated hydrostatic pressure.
7. Where mud or other sediment has been deposited on top of footings to be waterproofed with bentonite, require removal by water jet to assure intimate contact between concrete and bentonite.
8. For projects in coastal areas or other areas with potentially impure water or soil, specify a bentonite product grade with salt-resistance and require a groundwater-salinity and contamination analysis to verify compatibility.
9. Verify that substrates are consolidated, smooth, and clean to assure intimate contact between concrete and bentonite.
10. Seal expansion joints with suitable expansion joint material prior to bentonite system application.
11. Specify strip bentonite waterstops to be installed in all applicable cast-in-place concrete joints in foundation walls and slabs. Also use these flexible waterstops around pipe and structural elements penetrating the plane of the waterproofing course.

13 Plaza waterproofing

Steadily rising land costs have created the need for more efficient use of building space. Waterproofed plazas and roof terraces (sometimes called waterproofed decks) are a popular solution. These spaces were once limited to underground storage areas, access tunnels, and parking garages. With the advent of air-conditioning, they became suitable for computer rooms, offices, and public-assembly rooms; occupancies now routinely located below plazas, planters, and sidewalks.

Waterproofed plaza systems differ greatly from roof systems in their performance criteria. In conventional roof systems, the weather-exposed membrane must resist ultraviolet degradation and extreme temperature variations. A waterproofed plaza membrane experiences far less temperature change and is permanently shielded from ultraviolet radiation. Its primary performance criterion is to resist continuous exposure to moisture. For earth-covered membranes, resistance to vegetation roof penetration and soil contaminants are additional performance criteria. Because a waterproofing membrane may be far less accessible than a roof membrane, its anticipated service life must approximate the service life of the building. Reroofing is an inconvenience; re-waterproofing is a relative economic disaster. It costs at least 10 times the cost of a roof tear-off to remove up to eight feet of earth filled with plantings or to jackhammer a concrete topping and haul it away.

The scope of this chapter is limited to basic waterproofed deck systems and their foregoing sub-categories. Related systems featuring coatings for combination waterproofing/traffic surfaces for parking garages and balconies are not covered. The thin, liquid-applied coatings and traffic-resistant applications "waterproofing" these systems can resist only minimal hydrostatic pressure. As a consequence, very few are suitable for use on slabs over enclosed spaces with rigorous water-resistant performance criteria. Garage roof decks, defined as uninsulated decks above unheated garages; sports facilities; and grandstands are also not covered. These decks must withstand pedestrian and vehicular traffic, and resist oil, gasoline, and de-icing salts. Vegetative green roofs are also beyond the scope of this manual.

A. Categories

Waterproofing on structural slabs over occupied space falls into two basic categories:

1. Plazas or terraces, located at or slightly above grade.
2. Earth-covered structural slabs.

Plaza/terrace systems come in two basic sub-categories, in which the differentiating feature is whether the membrane is accessible or inaccessible. In the accessible sub-category, the system has a removable wearing course: pavers on insulation, pedestals, or sand beds. In the inaccessible sub-category are two further subdivisions: (1) systems with concrete protection slabs, and (2) systems with solid mortar-setting beds. Both subdivisions are classified as inaccessible because they require demolition – e.g., jackhammering of the concrete protective slab or of solidly grouted pavers – to provide access to the membrane. Removable pavers on removable insulation, pedestals, or sand beds are an exception.

Ruggeiro makes a further distinction within the plaza/terrace category. Plazas, or promenades, differ from terraces because they:

- are located at or near grade
- are almost always insulated
- have structural capacity for vehicular traffic (e.g., fire engines, snow-removal equipment)[1]

Earth-covered waterproofing systems are generally classified as vegetative green roofs. Membrane selection is mostly or entirely the same as those for plazas. Their special concerns of hydrology, soil contaminants, wind uplift, and root penetration are explained in many publications and are being addressed by an ASTM subcommittee.

A waterproofed plaza system with a separate wearing course contains (from bottom to top) some or all of the following components (see Figure 13.1):

- structural deck
- membrane
- protection boards
- drainage course
- thermal insulation
- concrete protection (working) slab
- flashing
- wearing surface

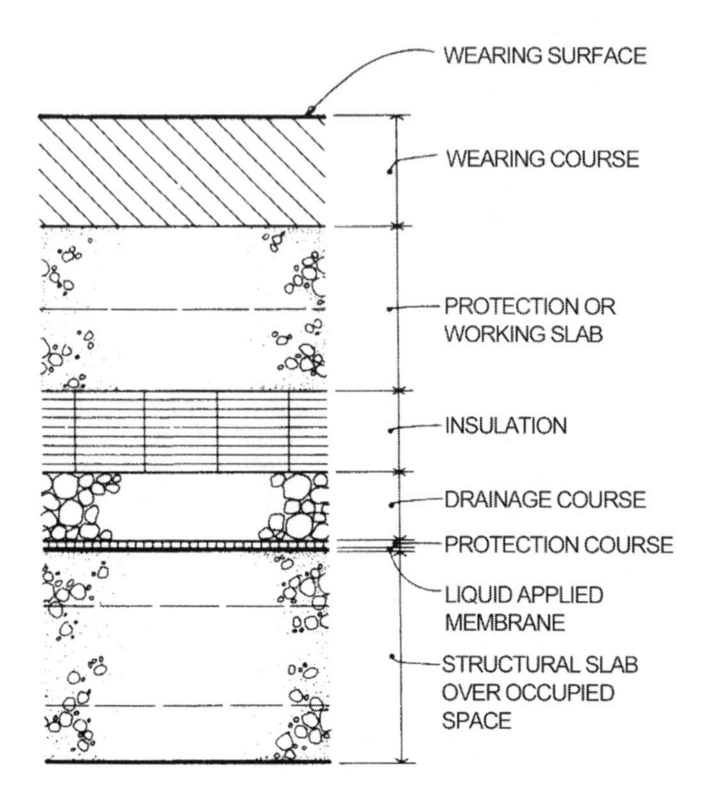

Figure 13.1 Basic components of membrane waterproofing over a framed structural slab. Insulation and drainage course are often reversed from the above order. (ASTM International)

Consult these ASTM standard guides:

- C981 for built-up membranes
- C898 for liquid-applied membranes, and
- C1127 for high solids content, cold liquid-applied elastomeric waterproofing membrane with an integral wearing course

Section B. incorporates material from these standards.

B. Structural decks

Cast-in-place, monolithic structural concrete slabs make the best substrates for waterproofed deck systems. Their continuity gives them a big advantage over precast concrete, which usually requires a concrete topping (normally two inches thick) to provide a smooth, continuous top surface uninterrupted by joints. The rotation of bearing ends of precast structural members can open and close joints, possibly requiring expansion joints to accommodate movement. Standard-weight concrete (150 pcf) is preferable to lightweight structural concrete (110-pcf minimum density), as it experiences less ultimate deflection from long-term creep. Lightweight insulating concrete is unsuitable as a deck substrate.

Dead loads maximize the effect of long-term creep on deflection. Preventing that necessitates a conservative approach to assure retention of adequate slope for drainage. To assure permanent retention of a one percent slope in a reinforced concrete slab, its initial slope should normally be *at least* two percent. Creep in structural concrete – i.e., increasing deflection under constant, prolonged loading – is inevitable in both reinforced and pre-stressed concrete. Total long-term deflection may equal three times instantaneous elastic deflection. Creep-prevention is an especially significant factor in structural concrete slabs supporting heavy overburden loads, as opposed to plazas, where intermittent live loads may constitute 50 percent or more of total design load. Structural design of plazas focuses more on live-loading. As an example of possible plaza live-loading, the service weight of a small, two-axle fire-fighting pumper can exceed 26,000 pounds in concentrated wheel loads.

You should make sure that the structural engineer is briefed about the need for a positive slope for drainage. Some supporting structures have been under-designed even for load-carrying capacity. Preliminary coordination among waterproofing designer architect and structural engineer is essential to assure adequate load capacity and slab slope.

A structural plaza deck typically consists of:

- reinforced concrete slabs;
- concrete topping over precast units;
- post-tensioned slabs; or
- composite concrete and steel decking.

Reinforced structural concrete slabs are the most common; they consist of framed or flat slabs. However, monolithic concrete slabs make a better substrate for waterproofing.

Joints between precast units and at their ends may not be in the same plane, and therefore require a concrete topping to even out lippage and cover lifting rings and welded plates. However, the topping is prone to cracking along joints and in particular at the end joints, which may rotate between precast elements at supporting girders.

Post-tensioned slabs offer better control of deflection and cracking within the plaza deck. Careful analysis of the deck deflection pattern and post-tensioning design is critical to achieve proper slope to drains after the plaza overburden has been placed.

Composite decking comprised of concrete on steel centering is more commonly used for terraces at higher floor levels than for plaza design near grade level. Provisions for venting moisture from the concrete must be made if a liquid membrane is applied to the concrete surface. This is typically achieved by using a slotted steel deck, or additional curing time, which can be reduced by a water/ratio of less than 0.4.

Regardless of which structural system you select, you should be certain that the structural engineer is informed of the need for positive slope to drain.

C. Membranes

Membranes for waterproofed deck systems include:

- conventional built-up bituminous;
- multi-ply modified-bitumen sheets;
- single-ply sheets; and
- liquid- or fluid-applied elastomers.

A built-up bituminous membrane consists of multiple plies of saturated felts layered between applications of bitumen applied on-site. The plies include glass mats or fabrics, and polyester mats or fabrics as reinforcement. The bitumen is unblown asphalt See ASTM D449, "Standard Specification for Asphalt Used in Dampproofing and Waterproofing."

Felt plies can be shingled or phased (ply-on-ply). In a phased type application, moisture that penetrates through a lap leads only to the next ply and not through the entire membrane. Membranes constructed of organic felt have not performed well when exposed to standing water. Glass fiber felts are less absorbent than organic felt.

Bitumen sheets modified with polymers have improved flexibility, elasticity, and the cohesive strength of bitumen. Some are self-adhering sheets, laminated to a high-density polyethylene backing and are often called "peel-and-stick" or rubberized asphalt. The sheets can be applied as a single or multi-ply waterproofing membrane system. They must be protected from ultraviolet exposure within a month of application (see Figure 1.3).

Modified-bitumen sheets used for roofing do not perform as well when used as waterproofing membranes because of the potential for wicking of the reinforcing at end laps. These systems are fully adhered to the concrete substrate and are sensitive to site conditions, moisture, and deck surface quality.

Single-ply sheets include ethylene propylene diene rubber (EPDM), butyl rubber, ketone ethylene ester (KEE), thermoplastic polyolefin (TPO), and PVC. Butyl rubber sheets have been largely withdrawn from the market. PVC and KEE sheets offer improved puncture resistance and heat-welded seams. These sheets are either fully adhered or loose laid. When loose laid, they should be compartmentalized by adhering the sheet in 10-foot grid.

This forms compartments that confine water migration if a leak occurs and also helps facilitate leak detection.

Liquid-applied membranes include hot and cold:

- polymer-modified asphalt;
- single blown asphalt;
- coal tar modified urethane;
- two-component urethanes;
- aliphatic polyurethanes;
- reinforced liquid polyester;
- two-component synthetic rubber; and
- polymer-modified asphaltic emulsion.

Solvated or emulsion-type membrane systems must have a minimum of 65 percent solids to reduce pinholing. Problems associated with reflective cracking from the deck below can be prevented by membranes with at least 60 dry mil thickness and by reinforcing them.

The advantage of adhered membrane systems is the localization of leaks. A disadvantage is the need for rigorous surface preparation of the substrate, which must be dry with a lightly broomed (sidewalk) texture, and dust-free. Cracks must be detailed. Liquid-applied membranes should never be used to fill or level surface irregularities.

Moisture is the adversary of these systems. It causes urethanes to foam and hot-applied systems to froth. Dust can cause pinholing or entrained air bubbles in the coating film. Unsuitable curing agents such as water glass can inhibit adhesion.

Hot-applied liquid systems are usually reinforced with polyester or woven glass. This process requires two applications, which minimizes coincident pinholing and thin spots in the membrane.

Cold-applied liquid membranes are applied over a concrete substrate by spray, squeegee, roller, brush, trowel, or other method recommended by the membrane manufacturer.

Manufacturers claim that these products are sufficiently elastic to bridge cracks that occur in the concrete after the coating is in place. Reflected cracking is reduced by increased thickness. The reinforced liquid polyester systems require exposure to ultraviolet light to cure. Polymethyl methacrylate (PMMA) cures from contact with moisture in the air. Urea based systems are fast drying and cure chemically.

ASTM International Standard C836 is a non-generic performance specification that describes the required properties and test methods for cold-applied elastomeric type waterproofing membranes for both one- and two-component systems.

Membranes for waterproofed deck systems include conventional built-up, modified-bitumen sheets, single-ply sheets, and cold and hot reinforced liquid-applied elastomers.

VOC limits, concern with fumes, and other safety issues have virtually eliminated coal tar pitch. Its few remaining uses are for repairing existing membranes. Currently, several materials are legal and can be used for limited patching.

As in roof construction, however, single-ply elastomers, weldable thermoplastics, and modified-bitumen sheets are rapidly replacing the more labor-intensive built-up membranes. A popular modified-bitumen sheet is a self-adhering single ply, laminated to a polyethylene backing. These peel-and-stick sheets are applied over a concrete substrate primed with either a solvent or an emulsion-type primer. They must be protected from ultraviolet degradation within a few days of application. See Chapter 9.A. for expanded discussion.

EPDM, PVC, TPO, and ethylene interpolymer (EIP sheets), with improved puncture resistance, are the membranes of choice today.

These single-ply elastomeric and thermoplastic sheets are either fully adhered or loose laid. When loose laid, PVC sheets are often adhered in grid strips, usually 10 feet each way, to form watertight compartments that aid in restricting leak-water migration and facilitating leak detection.

Liquid-applied membranes include hot and cold polymer-modified asphalt, single-component asphalt or coal tar-extended urethane, two-component urethane elastomer, PMMA and various polyether and polyurea systems. See Chapter 11.C.

Liquid-applied membranes' advantages include localization of leaks (compared with compartmentalization of leaks with loose-laid sheet membranes). Weighing against this advantage, however, is the need for extremely rigorous preparation of the substrate, which must be dry and largely dust-free, with patched cracks, as discussed in Chapter 11.C.

Hot-applied rubberized asphalts, heated in jacketed kettles (melters), are usually reinforced with polyester or woven glass. This process requires two applications to create a thicker, more stable membrane to minimize pinholing and thin spots. Since they cure as soon as they cool, these coatings are less vulnerable to rain or sudden temperature drops, which jeopardize the slower-curing cold-applied coatings. However, they are still susceptible to moisture in the substrate and must be applied to dry surfaces.

All of these highly elastic, liquid-applied coatings are capable of bridging small cracks. They have low permeability, are easily applied to contoured surfaces, and are generally self-flashing. Their chief liability is their great dependence on rigorous field supervision and application, requisites for uniform thickness, and film integrity.

Although spray applied LAMs increase productivity, their use creates high risks from hoisting spray equipment, overspray, high winds, and blowing dust and debris. Extra precautions are necessary.

For more detailed discussion of these various waterproofing membranes, refer to Chapter 6 "Positive-side Waterproofing" and Chapters 9 through 11. Remember the general rule of waterproofing design: avoid use of different, possibly incompatible materials when they require a joint to produce a continuous waterproofing film. When foundation walls also require waterproofing in conjunction with a plaza, choose the same waterproofing membrane material to facilitate the junction at the slab-wall intersection, if possible (see Figure 13.2).

D. Protection boards

All waterproofing membranes require shielding from construction damage and ultraviolet radiation as soon as practicable after installation and testing. The most common material is an asphalt-core, laminated panel, $^1/_{16}$, $^1/_8$, or $^1/_4$ inches thick, faced with polyethylene film on one side, to prevent sticking. Though sometimes used instead of protection boards, prefabricated plastic drainage composites should be used in addition to protection boards to prevent damage to the membrane before prior to installation and to ensure it isn't dislodged by wind before overburden is installed.

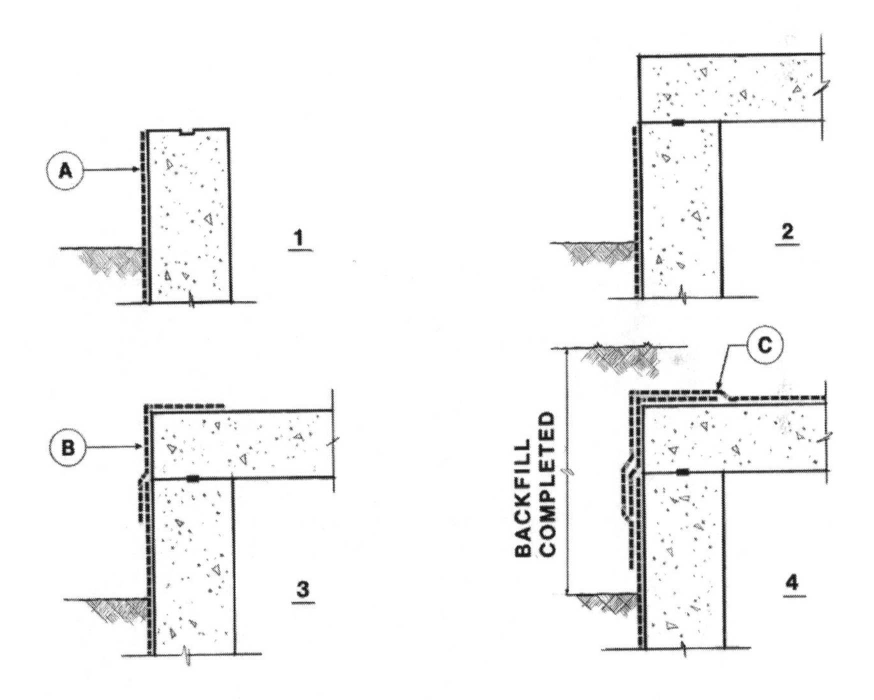

Figure 13.2 Details illustrating construction sequence, steps 1 through 4, show correct membrane lapping required to assure dependable waterproofing at the vulnerable joint between the foundation wall and the structural slab, a common source of waterproofing problems. If membrane material C is different from B, or B is different from A, impaired adhesion between membranes can promote leakage through the wall-slab construction joint

Installation of protection boards is mandatory immediately after completion of a horizontal waterproofing membrane and termination of integrity testing. Membrane application, testing, and protection-board installation must proceed in uninterrupted sequence, with no intervening delays exposing the membrane to the multiple hazards of other construction operations. Common membrane-puncturing hazards include:

- pipe scaffolds, sometimes erected without the required wood planks to distribute the concentrated loads (see Figure 13.3);
- stockpiled masonry units;
- reinforcing bars;
- welding rods;
- sharp-pointed fasteners;
- loose aggregate;
- debris scattered randomly over an unshielded membrane;
- contaminated oil dripping from pipe cutters; and
- mortar dropping from masonry construction underway at the perimeter.

Any of these hazards can end a membrane's service life before it has had a chance to prove durability. Delayed installation of protective insulation, drainage composite, and even a concrete protection slab is pointless if a membrane is already destroyed.

Figure 13.3 Unprotected membrane on a plaza over an auditorium

Field inspection is critical at this stage of a waterproofing project. Make sure that the client and manufacturer have qualified representatives at the site to assure that protection-board installation takes place immediately after the membrane has passed the integrity test. If the membrane fails the test, it must be repaired and retested.

E. Drainage design

A membrane waterproofing system is applied directly to a structural deck, and then covered with a wearing surface or overburden, when water is expected to reach the membrane. According to ASTM C981, drainage of the waterproofed deck system should include all components, from the wearing surface down to the membrane. The structural deck and the supporting columns and walls should be properly designed to provide positive slope. Inadequate slope to drain is a common deficiency in plaza design.

Drainage at the membrane level is required to:

- avoid building up hydrostatic pressure due to collected water against the membrane
- avoid freeze-thaw cycling of trapped water that could heave and disrupt the wearing surface
- minimize the harmful effect that standing, undrained water may have on the wearing surface material and membrane
- minimize thermal inefficiency of wet insulation or water below the insulation

A two percent 1/4-inch-per-foot slope is recommended for positive drainage. The substrate should slope away from expansion joints and walls. Gravel or plastic drainage panels can provide the necessary medium to facilitate water flow to drains, as can grooved or ribbed insulation boards.

Waterproofed decks should incorporate multilevel drains capable of draining all layers. These drains must permit differential movement between the strainer located at the wearing level and the drain body that is cast into the structural concrete slab to prevent shearing.

The drainage of a waterproofing system at the wearing surface level can be accomplished through an open- or closed-joint system. The open-joint system allows rainwater to quickly filter down to the membrane level and subsurface drainage system. A closed-joint system is designed to remove most of the rainwater rapidly by sloping surface drains that allow a minor portion to gradually infiltrate down to the membrane level.

Open-joint systems include pavers on pedestals or pavers placed directly on ribbed polystyrene insulation boards. Joints should be less than 6 mm (1/4-inch) wide to minimize catching high heels and cigarettes. Advantages are:

- elimination of the cost and maintenance of sealant joints;
- easy adaptability to a dead level wearing surface;
- faster and more efficient drainage; and
- easier access for cleaning and repairs to subsurface components.

The disadvantages are:

- rocking of improperly set pavers due to pedestrian traffic;
- unpleasant reverberations from heel impact; and
- possible hazards for pedestrians wearing high-heeled shoes.

Closed-joint systems consist of either a mortar-setting bed or caulked joints. This type of construction changes the waterproofing membrane to a secondary line of waterproofing defense, since the majority of rainwater is drained from the wearing surface level. The closed-joint system should slope away from adjoining walls and expansion joints to direct water away from above and below the wearing surface level. The advantages of a closed-joint system include:

- protection of the membrane from de-icing chemicals, dirt, and debris;
- ability to use a greater variety of paver types, designs, and sizes; and
- the comfortable feeling of a solid surface for pedestrians.

The main disadvantages are:

- extremely slow drainage; and
- imposition of a hydrostatic head of pressure on the membrane.

A system that uses brick or stone pavers in a sand-setting bed is better than a closed-joint system, but not as good as an open-joint system.

Insulation

The selection and location of insulation in the system is influenced by:

- deck design
- environment
- its physical and chemical properties
- characteristics of the wearing surface
- loads to be supported

Insulation placed over the waterproofing membrane and the protection and drainage layer provides the best results. It insulates the deck and membrane against extreme temperature cycles and the membrane can then function as a vapor retarder. Location of the insulation above the membrane also provides additional protection for the membrane.

The choice of insulation type is limited to extruded polystyrene (XPS) per ASTM International C578 Standard Specification for Rigid, Cellular Polystyrene Thermal Insulation. It must be able to accommodate the plaza dead and live loads; be dimensionally stable and compatible with the waterproofing membrane. It must also be as non-absorbent and resistant to freeze-thaw deterioration.

Drainage of a waterproofed deck system should incorporate all components, from wearing surface down to membrane, according to ASTM C981. Plaza drainage should minimize saturation of the wearing course, which might disintegrate from freeze-thaw cycling.

At membrane level, drainage is required to avoid the following:

- hydrostatic pressure from accumulated drain water
- freeze-thaw cycling of trapped water
- reduction of the insulation's thermal resistance

An absolute minimum of one percent slope, and preferably two percent, is recommended to assure positive drainage. Section B. "Structural Deck" explains the effect of long term concrete creep on slopes.

Either a percolating gravel stratum or plastic drainage panels can provide the pervious medium facilitating water flow to drains. In very cold climates, where melted snow flows over the membrane, siting the drainage course above the insulation may reduce the probability of condensation below the membrane. Conversely, this location may impair drainage, which is promoted by a below-insulation location of the drainage course.

You should specify multilevel drains, designed to permit differential movement between the strainer at the wearing course and the drain body embedded in the structural slab. A detail allowing relative movement between wearing course and structural slab prevents rupture of the drain body or connected pipes (see Figure 13.4).

At the wearing surface, drainage is accomplished via (a) an open-joint system that rapidly filters drain water down to the membrane level, or (b) a closed-joint system sloped to surface drains. Open-joint systems include pavers that are on pedestals or placed directly on ribbed insulation boards. Closed-joint systems have either a mortar-setting bed or caulked joints. An intermediate compromise is provided by bricks or stones in a sand bed. This compromise is inferior to an open-joint system, but superior to a mortar-set system.

F. Drainage courses

Specification of a drainage course above the membrane was one of several significant changes from design practices based on old coal tar pitch membranes and uninsulated systems. The lack of a slope to drain and water-pervious course to permit water to flow to it was responsible for

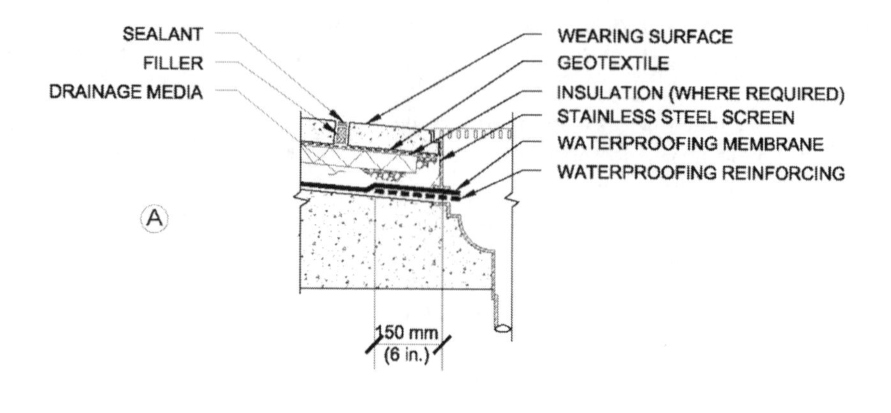

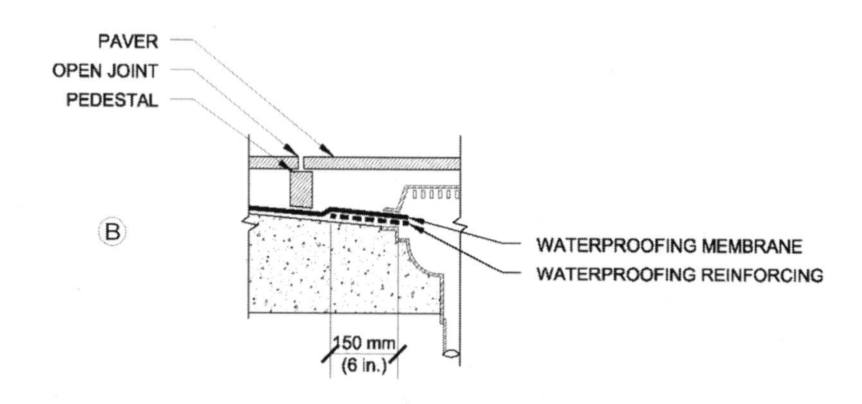

Figure 13.4 (a) Plaza drain with concrete wearing surface. (b) Plaza drain with pedestal supported pavers. Drain may be cast in concrete, welded to a sump or installed in a sleeve and equipped with underdeck clamps

numerous plaza waterproofing failures. A drainage course can comprise traditional stone aggregate, normally pea gravel or a drainage composite covered with a synthetic geotextile material. Geotextiles, made of polypropylene, polyester, or nylon fabrics, can be manufactured to a narrow range of permeabilities. They also serve a secondary function: resisting root intrusion down to membrane level in earth-covered waterproofing systems. They cannot, however, compete with a concrete protection slab as a virtually foolproof shield against root intrusion. See ASTM D7492 "Standard Guide for Use of Drainage System Media with Waterproofing Systems" for an exhaustive discussion of drainage media.

G. Thermal insulation

Until the energy crisis of the early 1970s, the key question was whether to specify insulation. Now, it is where to place the insulation: above or below the membrane. The old practice of simply relying on the aggregate or earth fill to insulate the occupied space is justifiable only when the space is neither heated nor cooled. Gravel, earth, and coarse sand provide high heat capacity, thus stabilizing heat gains or losses. But their thermal resistances (R values) are very low compared with those of insulating materials of equal thickness.

A third possible location for the insulation, under the deck, has several disadvantages:

- thermal bridging at walls and hangers
- subjection of membrane and deck to extreme daily and seasonable temperature cycling
- possible condensation of upward-migrating water vapor in the insulation
- fire resistance is required

The protected membrane roof (PMR) concept is generally the better design for waterproofed decks, for several reasons. Insulation located above the membrane:

- should maintain membrane temperature well above the dew point even in the coldest winter weather. This is a safeguard against condensation dripping down through cracks in the concrete slab.
- interposes the insulation between the potentially heavily trafficked plaza wearing surface and the membrane, thus augmenting the protection-board's membrane-shielding function.
- already requires high compressive strength, because the insulation in a waterproofed deck system normally carries higher traffic loads than a roof system does.

The above-membrane location for insulation in waterproofed deck construction limits the material choice to essentially the same as that for a PMR. As a practical matter, that choice is virtually limited to extruded polystyrene board, the only material possessing the two required properties of adequate compressive strength and moisture resistance. Foamglas® is the only other material satisfying both the compressive strength and moisture resistance required for below-membrane use. But its vulnerability to freeze-thaw cycling makes it too risky as above-membrane insulation, except in climates without freezing weather. Even in those climates, Foamglas® is limited to unfaced blocks rather than boards with two-sided facers. These kraft-paper facers lack the moisture-absorptive resistance of Foamglas® itself, and this deficiency disqualifies them as suitable insulation for a waterproofing system. Moreover, their brittleness makes them vulnerable to chipping of corners, cracking over irregular substrates, and other field hazards better resisted by the tougher extruded polystyrene.

Extruded polystyrene board should conform to ASTM C578 Type VII, a closed-cell foam that retains about 80 percent of its dry thermal resistance in a continuously wet environment. This foam's 60-psi compressive strength is adequate for any plaza or earth-covered design loading. Its roughly 10 percent permanent deformation under constant compressive stress of this magnitude results in a mere 1/4-inch lowering of the wearing surface with 2 1/2-inch-thick insulation. That will usually be insignificant.

For expanded discussion of the design factors involved in plaza waterproofing thermal design, see ASTM Standard Guide C981, "Standard Guide for Design of Built-up Bituminous Membrane Waterproofing Systems for Building Decks," Section 13, "Insulation." In addition to the design factors briefly discussed in this chapter, ASTM C981 covers dimensional stability, fungus resistance, vapor permeance, shear stress, fatigue stress, and several other design criteria. ASTM C981 also explains the problems of below-deck insulation, vapor retarders, and other aspects of plaza waterproofing design.

H. Concrete protection slabs

A concrete protection slab, at least three inches thick and normally reinforced with welded wire fabric, is an optional component in waterproofing systems for both plazas and earth-covered slabs. Its chief function in a plaza system is to protect the membrane during subsequent construction. Sometimes called a "working" slab, it provides a hard, stable surface for construction operations. In earth-covered systems it serves a permanent function, shielding the membrane from possible damage from roots that may penetrate the drainage course and protection.

Whether to include or omit a concrete protection slab is a major design decision in a plaza waterproofing system. You have to weigh accessibility against reliability. Accessibility to the membrane, provided by specifying removable components above the membrane, is an enormous advantage, facilitating repair or, in extreme instances, replacement of the waterproofing membrane. That benefit usually results in reduced reliability. High-density polyethylene and drainage courses with geotextile filters can provide considerable resistance to root intrusion into the membrane. They cannot, however, compete with the virtually impregnable root intrusion resistance of a concrete protection slab or provide protection from scaffolding and landscape equipment. Is this extra insurance worth the premium cost of the concrete protection slab? Does the reduced risk of failure assured by the concrete protection slab compensate for the higher cost of repair in event of failure? Can the system accommodate the additional 38-psf dead load and the additional 3-inch thickness of the concrete protection slab? You must answer these questions with a business decision based on conditions, surfacing, and sensitivity of the waterproofed occupancy.

I. Flashing

As with roof systems, waterproofed deck systems require flashing at terminations, penetrations, and expansion joints. In contrast with roof systems, where base flashing installation follows membrane installation, flashing at waterproofed deck terminations must be installed prior to membrane application. In the plaza field, at rising walls and expansion joints, the horizontal leg of base flashing should be installed on a curb at least 1½ inches above the structural slab to direct water flow away from the joint. See Appendix C.

Internal and exterior corners and similar transitions must be reinforced with at least one membrane ply or liquid-applied fillet. Raising this flashing height above the membrane is good practice, like the similar practice for expansion joints.

Penetrations are flashed like roof penetrations. Individual pipes should be spaced a minimum of six inches apart. Avoid ganged pipes in pitch pockets.

Base flashing must extend above the wearing course, preferably eight inches. This is critical if the wearing course is a closed system.

Like flashing in roof systems, flashed penetrations at walls and expansion joints should always be at high points, with the drained surfaces sloped away from them. If practicable, penetrations should also be located at high points, but, in any event, never at low points.

At doors opening onto plazas where ADA governs the transition between the plaza and the interior, a gutter with ADA grating will capture the water flowing down the façade, prevent it from driving under the sill and solve the height differential problem (see Figure 13.5).

J. Expansion joints

Large plazas may also require structural expansion joints in the plaza field, generally spaced at 100 to 200 feet. These joints are usually designed by the structural engineer. Wearing courses, which are exposed to temperature changes 10 or more times as great as those experienced by the insulated, underlying structure, require a correspondingly greater number of expansion joints extending only through the wearing course section. A wearing course expansion joint must coincide with each structural expansion joint. At these locations, the expansion joint width at the wearing course must approximate the width of the structural expansion joint. The only exception to this rule is where the wearing course consists of loose-laid units. See Appendix E.

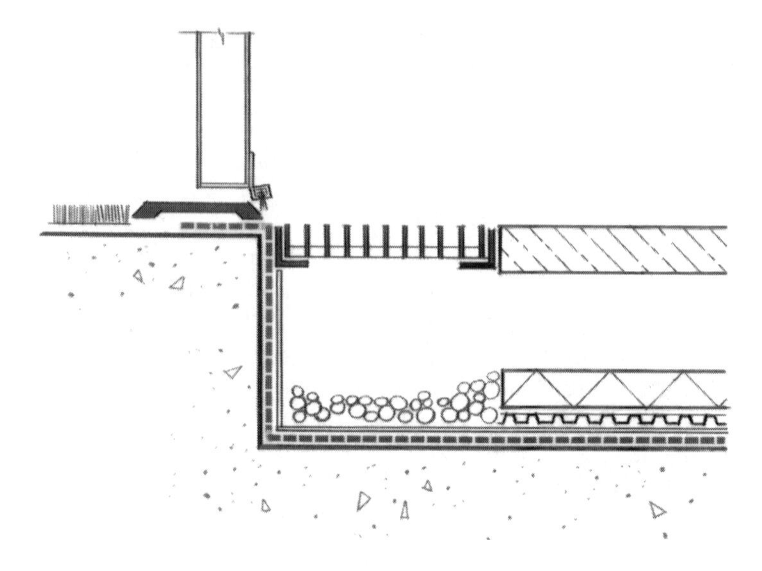

Figure 13.5 This detail shows a grating-covered gutter at a door opening at the same elevation as a pedestal-supported, paver-surfaced plaza. It satisfies ADA requirements and prevents wind-driven rain from penetrating under the door or threshold

K. Wearing-surface design

Wearing surfaces are divided by ASTM C898 into two basic categories:

1. Open-joint systems, drained at the membrane level.
2. Closed-joint systems, drained at the surface.

The membranes of open-joint systems, which resemble PMR roof systems, are rarely subjected to hydrostatic pressures exceeding 5 psf (i.e., a 1 inch depth of water). Closed-joint (i.e., monolithic) wearing surfaces, designed to shed water at the surface, similarly shield the membrane from continuous high hydrostatic pressure.

In heavily trafficked locations, waterproofed plazas must satisfy these wearing-surface design criteria:

- structural strength to bear traffic loads
- durability under heavy wear and weathering

You may also consider:

- aesthetic appearance (on plazas and roof terraces)
- heat reflectivity (to avoid summer temperature buildup)

Cast-in-place or precast concrete, ceramic, or masonry units generally satisfy these requirements. Dark colors – especially black – are normally avoided because of their high heat absorption.

Joints on a roughly 10-foot-maximum grid are normally required to accommodate thermal contraction and expansion in surfacing units. They can experience daily temperature extremes of 80°F and annual extremes of 180°F in the most severe climates. Use of snow-melting equipment (embedded electrical-resistance cables or hot-water pipes) complicates the thermal design of the surfacing. Rapid temperature changes accompanying intermittent operation of the snow-melting system or expansion of corroding embedded pipes can crack surfacing slabs. Heated snow-melting equipment that may be embedded in concrete or over membranes may soften bituminous materials and impair their performance.

L. Open- versus closed-joint plazas

The first step in surfacing design is choosing between open and closed joints in the plaza surfacing. If the plaza must carry heavy vehicular live loads (for maintenance, fire-fighting, and snow-removal equipment) this preliminary decision may eliminate open-joint construction, which can seldom carry concentrated wheel loads of 7,000 pounds or more. Pedestal-supported pavers are especially vulnerable to such concentrated loads. When open-joint construction can be considered, however, it offers several advantages (listed later) over the closed-joint construction dictated by high vehicular loading.

Pavers for open-joint construction are fabricated from two basic materials: precast concrete and natural stone. Precast pavers are hydro-pressed or wet-cast in molds into units of square or rectangular shape, ranging from 12 by 24 inches to 30 by 30 inches, in thicknesses normally ranging from 1 $^{1}/_{4}$ to 2 $^{1}/_{2}$ inches, but sometimes as thick as 4 inches. These pavers normally weigh 15 to 25 psf. Compressive strength is generally 7,000 psi; maximum water absorption 5 percent. Precast pavers are available in a wide spectrum of colors and textures. (For detailed data on all types of pavers, see Appendix B.)

Natural stone pavers include granite, marble (including travertine), flagging of slate, and high-density limestone. Thicknesses range from 1 $^{1}/_{2}$ to 4 inches, for stones with low flexural strength. Shapes and sizes vary, but for precast pavers, rectangular and square shapes are preferred for pedestal-supported natural stones.

Pavers for closed-joint construction cover a wider range than for open-joint construction. In addition to all the precast and natural stones used in open-joint construction, closed-joint pavers can include bricks, concrete masonry units, and ceramic and quarry tile. Paver bricks should never be facing brick.

Open-joint surfacing units – precast concrete, tile, or masonry units – can be installed on ribbed insulation that conducts water downward to the waterproofing membrane, where it runs to drains. Fine gravel, clean coarse sand (held on a No. 30 sieve), or no-fines (porous) concrete can serve as the percolating stratum underlying the open-joint surfacing units. Loose aggregate can accommodate freeze-thaw cycling of entrapped water, with freezing expansion contained harmlessly within the aggregate interstices.

In a frequently specified and generally superior alternative to loose aggregate percolation, the surfacing units can be installed on pedestals, on insulation, or directly on protection boards; providing faster, less obstructed subsurface drainage. Surfacing soffits can also be grooved, corrugated, or otherwise contoured to create open paths for subsurface drainage.

Filter cloth is not required under open-joint pavers for two reasons:

1. It retards drainage (by retaining dirt), although it may reduce clogging in insulation board joints.
2. Filter-cloth materials, polyester and polypropylene, degrade from ultraviolet radiation that penetrates the open joints.

This alternative pedestal-supported, open-joint system has several advantages:

- adaptability to a dead level wearing surface;
- faster, more efficient drainage, with better surface-dirt removal;
- ventilation drying of subsurface areas;
- easier access for maintenance, cleaning, and repair of subsurface components; and
- elimination of joint sealants required in closed-joint systems.

Offsetting these benefits, open-joint plaza construction has several liabilities:

- a more complex design/construction problem, at probably higher initial cost;
- possible hazards for pedestrians wearing spiked heels (on plazas with wide joints between pavers);
- rocking of incorrectly set pedestals under pedestrian traffic; and
- objectionable reverberations from leather heel impact.

Monolithic (closed-joint) construction changes the waterproofed deck design philosophy, demoting the membrane from the primary to the secondary line of waterproofing defense. The membrane nonetheless is an essential component of the system because the surfaces of waterproofed decks are normally exposed to traffic loads, drastic temperature changes, and freeze-thaw cycling, with expansion of water trapped in cracks, holes, or other imperfections. Insulation is interposed between the membrane and the wearing surface material, with a setting bed for the surfacing units: precast ceramic or masonry tiles or manufactured conglomerates. Surfacing should slope away from adjoining walls and expansion joints to direct water away from them both at and below the surfacing.

For closed-joint construction, there are three methods of paver support: sand beds, mortar beds, and direct bearing on the insulation. Sand beds work best with small pavers because they can accommodate movement of the paver units in the vertical plane with less differential elevation between adjacent units. (Where adjacent units have different elevations adjacent to joints, they may trip pedestrians.)

Sand-setting beds can be installed over a concrete protection slab but are generally unsatisfactory on rigid insulation. Sand-setting beds should be restricted to narrow thickness to avoid paver settling, rocking, and loosening caused by eccentric loading. Unbalanced compressive stresses in the sand can cause differential displacement in the sand stratum, with consequent tilting of the supported pavers. To avoid these paver disruptions, sand thickness should be limited to one inch, which is usually adequate to accommodate subgrade variations and a higher substrate elevation at expansion joints.[2] A one-inch sand bed can be installed over filter cloth on a three- to six-inch stratum of coarse aggregate or drainage composite. This practice provides the additional benefit of preventing frost heave in climates subject to freeze-thaw cycling. Sand should be coarse, with single-sized grains – not gap-graded like mortar sand – to promote maximum porosity.

Closed-joint construction offers additional advantages:

- protection of waterproofing membrane from de-icing salts, cigarettes, leaf debris, and dirt in general;
- adaptability to a greater variety of pavers, designs, and sizes;
- shielding of insulation and waterproofing from large quantities of water;
- enhancement of pedestrians' sense of solidity; and
- damping of cacophonous clatter of high heels striking pedestal-supported pavers.

One minor disadvantage is that it drains slowly and imposes a hydrostatic pressure on the membrane.

M. Plaza structures

Planters, reflecting pools, foundations, and other plaza "furniture" should be installed above the waterproofing membrane to maintain continuity of the membrane and avoid impeding drainage (see Figures 13.6 and 13.7).

N. Drains (also see Chapter 14.D.)

Two-stage or two-level drains, sometimes called "all-level," are mandatory for structural plaza slabs. The upper drain must be isolated from the lower

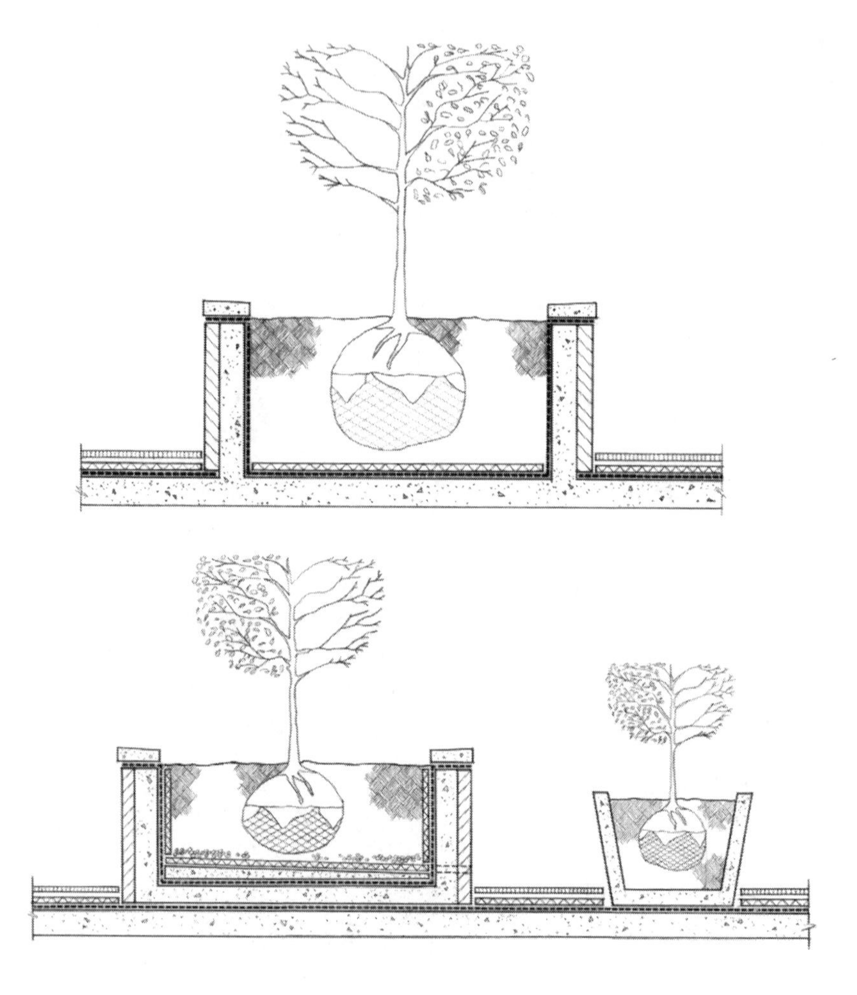

Figures 13.6 and 13.7 Show details of waterproofing for tree planters. The waterproofing shown in Figure 13.6 is interrupted by the walls of the planter, which is not as effective as the continuous waterproofing shown in Figure 13.7. Figure 13.7 shows the preferred detail, with a separate waterproofed container for a tree planting. Insulation is often installed in the planter to avoid false blooming in winter

drain to permit differential movement. (Wearing courses on plaza decks move independently of the structural slab.) To prevent damage to the drainage system or ruptured flashing from this movement, drains should be installed in the structural slab with gratings in the wearing course isolated from the drain body.

Do not use a drain unit with drainage at the membrane level provided by weep holes. Weep holes have a tendency to clog from crystallized soluble salts in the concrete or from coal tar pitch.

Lead reinforcing is recommended for membranes where pipes or drains are installed in sleeves, but is unnecessary where they are cast into the wall or slab. Because of their rigidity and durability, cast-iron drains, cast into the slab with flanges flush with, or slightly below, the slab surface perform better than aluminum, plastic, or sheet metal drains.

A minimum of one additional ply of membrane should be installed at drains. Drain-strainer frames should not be designed to support wearing courses. Lead, if used, should be carried into the drain bowl. Built-up and single-ply membranes and reinforcing should lap drain flanges by four inches and be secured with clamping rings. LAMs are applied over drains with extended flanges. Clamping rings are not used. Specify sediment buckets where wearing surfaces are water-permeable and where debris is expected.

Root intrusion is a common cause of clogged drains in earth-covered suspended slabs. Access for essential periodic maintenance can be provided by a perforated pipe and cleanout plug at the surface or a clay or concrete pipe manhole fed by subsurface perforated pipes installed in a drainage medium. The top is buried below the surface and identified with a telltale.

O. Testing

A failed waterproofing system is more destructive and expensive to correct than a failed roof. The replacement cost of a failed roof can be anywhere between $15 to $22 psf, a failed waterproofed plaza between $75 and $125 psf or more.

Therefore testing a waterproofed deck after the flashing and membrane have been installed is advisable. ASTM D5957 "Standard Guide for Flood Testing Horizontal Waterproofing Installations" provides the method for testing the water-tightness of a waterproofed deck. Some requirements set by this standard are:

- Slope of deck or membrane to be tested must not exceed two percent (1/4-inch) per foot.
- Membranes must be LAMs, adhered or loosed-laid sheets, or built-up and modified membranes.
- Do not test until 24 hours after the membrane has been installed (this requirement increases to 48 hours if the membrane was installed at ambient temperatures below 50° F).
- Inspect and repair flashing and membrane prior to testing.

ASTM D5957 says if a leak occurs during testing:

- drain water;
- locate and repair leak; and then
- re-test area under the same initial conditions.

Flood testing is tedious and expensive. An observer must be continuously present under the deck being tested to determine if a leak occurs. Then the test must be immediately terminated, the leak repaired and the membrane retested. Contrast this with the ease and speed of testing electronically with electric field vector mapping.

Electric field vector mapping (EFVM™) is a tool for improving quality control of waterproofing systems. Although relatively new to the U.S., it has achieved a long record of success in Europe. The system was pioneered in Germany. EFVM, unlike other leak detection methods, can quickly and accurately locate the point of water entry (see Figure 13.8).

EFVM uses water as an electrically conductive medium. A wire loop is installed around the perimeter of the area to be tested and introduces an electrical potential. The area within the loop is dampened and forms the upper electrical plate. The structural deck then becomes the lower plate.

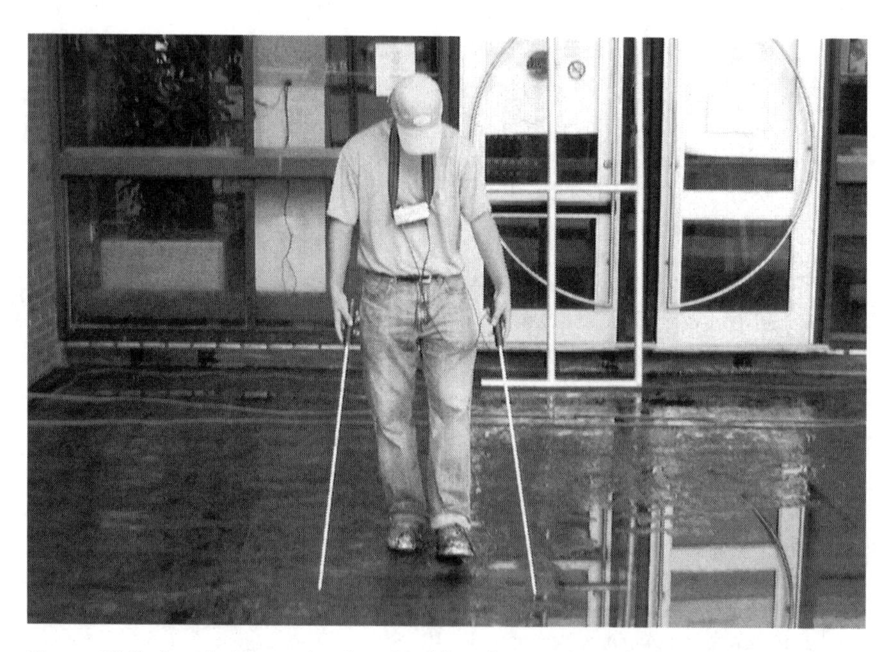

Figure 13.8 A technician using hand-held probes connected to a voltmeter to locate leaks in a self-adhering rubberized asphalt membrane in the student union plaza over occupied space at Rutgers University. A voltage potential between the membrane and the deck is established and the probes are used to track leakage current passing through the wetted membrane

The membrane acts as separator and insulator between the two plates. If moisture enters a defect in the membrane, electrical contact is established. The survey technician can then follow the direction of the electric field to the membrane defect. Advocates of EFVM state that the test method:

- locates defects precisely, enabling efficient repairs
- is able to re-test repairs immediately
- can be used after cover systems are installed, especially with green-roof landscapes
- is less expensive than conventional flood testing
- eliminates the hazard of overloading structural decks during testing
- can be used on steeply sloped roof surfaces where flood testing is impossible

The suitability of EFVM depends on the electrical resistance of the waterproofing materials and not all membranes may be compatible with this test method.

P. Design checklist

1. Provide a minimum 1/8-inch slope (preferably 1/4-inch) in a structural deck.
2. Slope deck away from walls and expansion joints.
3. Specify all-level drains for closed-joint waterproofed deck systems.
4. Provide a-3000 psi concrete topping, at least three inches thick, reinforced with welded wire fabric over precast concrete decks. Precast concrete units must be secured together (to prevent relative movement and consequent cracking of the topping).
5. Obtain the waterproofing manufacturer's signed approval confirming the suitability of the product for a specific project. Also require manufacturer's certification of the applicator.
6. Require the concrete contractor, not the roofing contractor, to repair defective concrete surfaces.
7. Maximize single-ply membrane sheet size (to minimize vulnerable field-spliced lap seams).
8. Reinforce substrate joints or cracks.
9. Compartmentalize loose laid single-ply systems if not fully adhered.
10. Locate expansion joints at slab high points.
11. Use two-stage drains.
12. Use protection slabs where rising walls, parapets, stairs, and trees will be installed over a membrane.

Notes

1 Steve Ruggerio, *Effective Plaza Deck Waterproofing Seminar*, University of Wisconsin, Dec. 16, 1994.
2 Estenssoror and Perenchino, "Failures of Exterior Plazas," *Construction Specifier*, Jan., 1991, p. 87–92.

14 Waterproofing failures

A. Causes of waterproofing failure

Waterproofing failures stem from the following causes:

- designer error
- negligent construction practices
- defective materials

Designer error is probably the chief source of waterproofing failure. Roof-system failures, in contrast, are chiefly attributable to contractor negligence. Designer error sometimes entails specification of an inappropriate system, through ignorance of its particular limitations and widespread confusion about the proper uses and limitations of various waterproofing systems. This simplest kind of designer error represents a substantial proportion of this source of waterproofing failures.

One explanation of the prevalence of design ignorance as a cause of waterproofing failure stems from the contrasting educational efforts of roofing and waterproofing manufacturers. After suffering along with architects, contractors, and building owners through a quarter century of rampant litigation, roofing manufacturers began providing reference details for a wide spectrum of project conditions. Most waterproofing manufacturers began providing technical guidance relatively recently. As a consequence, waterproofing designers have sometimes made disastrous mistakes, such as attempting to combine incompatible materials for walls and structural slabs with no satisfactory detail at their intersection.

The greatest source of designer errors may be a lack of coordination among the professionals involved in waterproofing design. Roof design is normally the sole responsibility of the architect (or a roof consultant retained by the architect or client). In waterproofing design, however, responsibility sometimes gets blurred because a soils engineer, a structural engineer, and an architect are involved in the design process. The necessary coordination sometimes gets lost in the general shuffling of the myriad problems and responsibilities involved in the construction process.

Consider a typical waterproofing project problem. The recommendation for waterproofing originates with the soils (geotechnical) engineer. The soils report also includes recommendations for foundation design: soil-bearing capacity, depth of footings, seismic forces (if any), etc. Following the soils report's recommendation, the architect and structural engineer jointly decide on the extent of required waterproofing. Plaza waterproofing is designed and detailed by the architect; waterproofing for basement slab-on-ground and foundation wall may be designed by the structural engineer and also by the architect. This duplication can doom the project.

As head of the design team, the architect is responsible for waterproofing design. The architect should provide details for all waterproofed components: plaza structural slab, foundation walls, and basement slab-on-ground. The engineer, however, may duplicate part of this work – e.g., indicating waterproofing for foundation walls and slab-on-ground with vague notes and lines. The engineer should refer to the architectural drawings for all waterproofing information, providing a unified source of information for the contractor.

Contradictory specifications can pose an even bigger problem, and a more important one since specifications usually outweigh drawings in court. The architect's specification may call for membrane waterproofing whereas the engineer's specification (following engineers' proclivities for these materials) may call for bentonite or negative-side coatings. The architect must always follow through, eliminating such conflicts in contract documents. When a conflict exists, the contractor may make a choice without consulting the architect. Poor communication among design-team members can be as deadly a foe of successful waterproofing projects as technical error.

Yet, technical errors do account for a large proportion of design-caused waterproofing failures. Defective design is one of the most common sources of waterproofing failure, exemplified by the following actual examples:

1. In a deep basement, part of the basement slab-on-ground is stabilized by rock anchors designed to resist hydrostatic uplift. The remaining slab-on-ground, not subject to this uplift pressure, is continuously supported on a gravel course with under drains. After completion of the basic design, two sump pits were added: one in the uplift zone, the other in the continuous bearing zone, with one wall of the sump pits on the same plane as the foundation wall. This unusual, if not unique, arrangement required an elaborate detail but was not brought to the attention of the designer. The contractor's "solution" was a dismal failure; it leaked.
2. Defective detailing caused a failure at the expansion joint in a waterproofed structural split slab over an underground gymnasium supporting a plaza. The defective detail, relying on the dubious principle of sliding metal plates, was squeezed into the same plane as the

horizontal membrane. It failed at a T joint and was virtually irreparable. Lesson: Expansion joints should be straight and extend vertically through the entire structural cross section.

A large variety of negligent contractor practices accounts for many waterproofing failures. Perhaps the least excusable is the omission of protection boards. Puncturing of unprotected membranes over waterproofed slabs is a common source of waterproofing failure. Erection of scaffolding, movement of heavy equipment, or damage from reinforcing bars and general construction debris all threaten a membrane unshielded from such hazards (see Figure 14.1). Such inexcusable incidents highlight the absolute necessity of inspection by a client's representative. Practically, if not legally, failure to provide such inspection makes the owner an accessory to the contractor's neglect.

The biggest villain among contractors involved with waterproofing projects is most likely to be the site contractor responsible for backfilling. Unacceptable backfill materials – rocks, frozen soil, organic matter, and miscellaneous debris – are often dumped into excavations regardless of specifications banning such materials. Some site contractors neither know

Figure 14.1 This chaotic scene, depicting an unprotected waterproofing membrane exposed to multiple punctures from reinforcing bars, wood, and nails, plus other construction debris, highlights the necessity of a client-financed inspector on the site to demand installation of protection boards

nor care that a bulldozer blade can be a weapon as well as an earth-moving tool. Inadequate filter-cloth lapping, sufficient to prevent infiltration of fines, or tearing of filter cloth during backfilling operations can compound these backfilling disasters.

Next in the villains' gallery is the general contractor who presses the waterproofing subcontractor into the premature installation of materials onto unsuitably prepared, or even unprepared, substrates. The general contractor obsessively dedicated to the proposition, "Time is money," can be the client's, and ultimately his own, worst enemy. Good workmanship outranks time-saving in waterproofing project economics. A tiny premium for waterproofing insurance will return a double or even triple-digit multiple over a penny-saving premature schedule accelerated by a reckless general contractor.

Other negligent practices include contractors' lack of preparation for the inevitable vicissitudes of construction. Under the best circumstances, delays resulting from weather or dewatering equipment failure can disrupt a waterproofing project. The contractor must nonetheless be prepared to cope. Installation of wall waterproofing sometimes proceeds so far ahead of backfilling that bad weather can cause long delays in backfilling operations, leaving the waterproofing exposed for weeks. Installing waterproofing on a basement mud mat while simultaneously excavating an adjacent area is another hazardous, yet not uncommon practice. The excavating operation can ruin the waterproofing application as dust and dirt are blown and tracked onto the finished waterproofing, or worse yet, onto the waterproofing as it is applied.

B. Compound errors

Identifying the cause or causes of a waterproofing failure may be more complicated than simply blaming one party. Construction culprits often have accessories. The designer, in particular, is likely to be a party to a complicated case involving waterproofing failure. In almost any litigation over a waterproofing failure, the designer's judgement can become an issue unless the contractor or materials manufacturer has displayed provable negligence.

Consider this complex waterproofing failure. Long steel girders, spanning a foundation excavation were dropped into bearing pockets and grouted solid. Solar heat raised the temperature of the exposed girders, and the consequent expansion cracked the foundation wall and its waterproofing. The contractor may be liable for prematurely grouting the girders and failing to anticipate a steep temperature rise, but the architect did not detail the expansion or alert the contractor to this hazard.

On another project, a structural slab-on-ground was cast over a waterproofed mud mat, and designed to resist hydrostatic uplift pressure. The contractor, however, erroneously deactivated the dewatering system before interior shear walls for stairs and elevator shafts were in place to resist the upward movement of the slab. Again, a lack of communication or coordination between designer and contractor produced a failure.

As a third example, a deep foundation wall was cast on a heavily trafficked avenue. The dynamic loads from trucks and buses caused the wall to deflect inward after it had been waterproofed. The vital lateral support of an intermediate structural slab in the deep, two-level basement had not been cast. The wall was later jacked back into proper vertical alignment and the structural slab providing its lateral support was cast. That was too late for the waterproofing on the foundation wall. It had split from the wall's excessive deflection. That might have been prevented if the designer and contractor had realized the necessity for temporary lateral support of the foundation.

As demonstrated by these examples, you must constantly picture the construction process to anticipate the fiendish ways that things can go wrong.

C. Foundation wall failures

Foundation failures can be attributed to:

- improper backfilling
- defective joint details
- concrete wall defects
- defective flashing at penetrations
- vapor entrapment
- abuse of the pre-applied membrane

Improper backfilling can cause failure via two mechanisms. The most obvious stems from inclusion of membrane-damaging materials within the backfill. Those include sharp-edged rocks, tree branches, or roots capable of puncturing protection board and ultimately penetrating the membrane.

Backfill settlement, a subtler, yet readily avoidable problem, also results from contractor negligence. Backfill settlement is an aggravating factor that compounds the perpetual gravitational slippage force to cause slippage. It can add the much greater frictional force of sinking soil mass in contact with protection boards, which then transmits the frictional force to the membrane. Even if the adhered membrane resists this magnified slippage force, the protection boards can be dragged downward, exposing the upper part of the membrane.

Backfill settlement results from negligently supervised backfill dumping. To avoid harmful settlement, backfill must be placed in small "lifts," a maximum of a foot or so (depending on soil characteristics), mechanically compacted to a density approximating its natural density. Proper backfilling depends upon a competent, responsible contractor, and even more so, on rigorous field inspection by the client's representative.

Defective joint details cause waterproofing failures at the tops and bottoms of foundation walls. At the top, where structural slabs support plazas or earth-covered planted sites, the horizontal construction joint where the structural slab bears on the foundation wall creates a plane of potential leakage, especially if not properly reinforced to prevent the joint from opening as the slab deflects or shrinks. The waterproofing membrane should be reinforced at this joint to prevent possible stress concentration and splitting of the membrane there (see Figure 14.2).

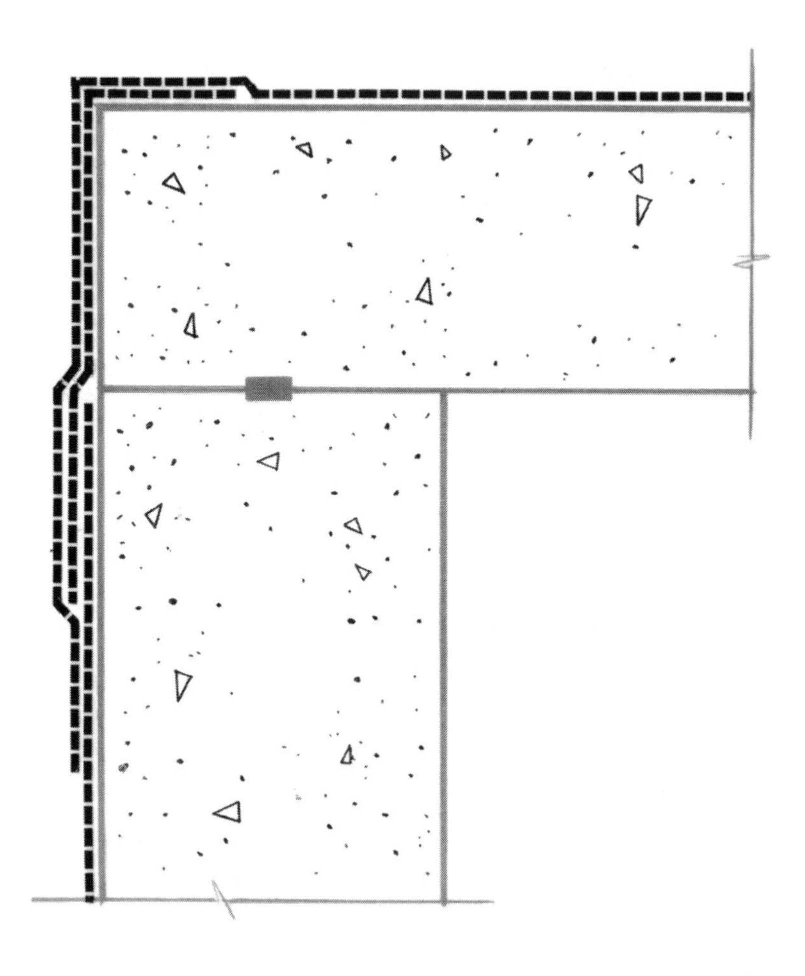

Figure 14.2 Where structural slabs bear on a foundation wall, it creates a potentially dynamic horizontal joint. Frequently, the foundation waterproofing is completed before the slab is cast. The membrane should be reinforced at this joint with a separate strip that may be combined with the corner reinforcing as shown in the above detail

At the bottom of foundation walls, a typical (and defective) detail is also vulnerable to leakage. Waterproofing carried across a wall footing at the same elevation as an adjoining mud slab creates a potential leakage path similar to the slab-wall joint described in the preceding paragraph. A vertical, positive-side waterproofing membrane is often extended down from the wall and then horizontally over the wall footing's toe. In the inevitable time gap between the application of the mud-slab waterproofing across the footing and the application of the overlapping wall membrane, the horizontal joint may be soiled by deposits left by excavation or storm water covering the wall footing. This soil residue can destroy adhesion between the underlying and overlapping membranes, leaving a defective horizontal joint vulnerable to leakage.

Failure at this wall-footing joint often occurs with bentonite waterproofing, where the joint is either poorly packed or not packed at all. (For rectification of this problem, see Figure 12.3.) Lack of temporary protection board to protect against foundation formwork installation and stripping also causes failures at wall-footing joints.

Concrete wall defects are another source of foundation wall problems. Unlike the surfaces of structural slabs and slabs-on-ground, which must be finished to a more or less continuous plane surface, the vertical surfaces of concrete walls are sometimes pocked with irregular indentations or tie-rod holes, honeycombing voids, rock pockets, formwork kickouts, or physically damaged areas. Hydrostatic pressure can produce concentrated shearing and flexural stresses that ultimately rupture a membrane adhered to an irregular concrete surface and admit leak water.

Contractors who fail to rigorously inspect concrete substrates destined to receive waterproofing before it is installed are negligent. Irregular concrete surfaces should be patched or repaired before membrane application is allowed to proceed. (See Chapter 16 "Specifications" for details about wall repair procedures.)

Defective flashing at penetrations for pipes and conduit sometimes occurs because of the designer's failure to require spacing at least six inches between adjacent penetrations.

To eliminate these problems, locate penetrations high enough to avoid interference with water stops and reinforcing bar splices. As a safe minimum dimension, penetrations should be located at least 12 inches above the top of the slab.

Vapor entrapment springs from an error in which the designer specifies membranes on both sides of a foundation wall. Entrapment of water vapor, given no chance to vent in either direction, can spawn a crop of blisters at the substrate-membrane interface, at consequent peril to a membrane's waterproofing integrity.

Abuse of pre-applied membranes is a more prevalent cause of waterproofing failures that can be traced to the lathers who install the waterproofing after the blindside membrane is in place. One form of abuse is the ill-advised use of torches to cut the rebars that often burn through the adjacent membrane. But a more frequent and egregious practice is to install braces or supports for the rebars. One common practice is to drive nails through the waterproofing to support horizontal bars (see Figure 14.3). A second is to use the membrane covered substrate as support for a rebar brace (see Figure 14.4). These penetrations of the membrane often go unobserved since the waterproofing installer and QA observer are long gone from the project when the lathers begin work.

D. Plaza slab failures

The major causes of plaza waterproofing failures are:

- lack of proper drainage
- omission of a drainage course
- failure to install protection boards immediately after completion of membrane installation and membrane integrity testing
- inadequate slope
- defective expansion joints
- membrane defects
- insufficiently dry substrate

Figure 14.3 Lathers used a nail driven through this pre-applied membrane to support the rebars. Damage from this common practice must be repaired before the concrete is cast

Figure 14.4 This HDPE blindside membrane has been burned when the lather burned the rebars with a cutting torch. Rebars should always be sawn off and torches prohibited

Lack of proper drainage has historically been a major cause of plaza failures. Failures at wearing surfaces were once common. Those failures were attributable to inadequate, single-level drainage designed to remove water from the surface, but not at the membrane below. The designer sometimes specified drains with weep holes at the membrane level, an effort that indicated a proper attempt to provide two-level drainage. But these weep holes often got clogged with bitumen moppings or cement leached from the pavers' setting bed. Drain water dammed in ever-widening areas around the clogged drains. This promoted efflorescence and consequent popping of tiles (from expansion of the grout). The ponded water also promoted mold and vegetation growth in paver joints.

Omission of a drainage course was also once common. Earth-covered slabs often require two drainage courses, one above and one below the insulation, to avoid drain clogging. A detailed explanation of how to avoid this problem is in Chapter 13.F. "Drainage courses."

A contractor's **failure to install protection boards immediately after completion of membrane installation and flood testing** is as risky as omitting them. The membrane-puncturing hazards of reckless backfilling operations threatening a foundation wall's waterproofing integrity have their counterparts on horizontal surfaces. Hazards include welding rods;

reinforcing bars; sharp-edged fasteners; loose aggregate; carelessly stock-piled masonry units; and pipe scaffolds, sometimes erected without wood plank supports to distribute concentrated loads (see Figure 14.5).

Inadequate slope in either of two locations often causes waterproofing failure. Slope should be designed to be at least one percent after dead-load creep and maximum elastic deflection. Plaza junctures with walls and expansion joints are common leak-prone locations Leakage usually results from the designer's failure to locate these flashings at high points, with the slab sloped away from them. Closed-joint wearing surfaces are the other common location.

Defective expansion joints are a common location for plaza waterproofing failures even without inadequate design for slope. Expansion joints in horizontal surfaces are especially liable to leak because of thermally induced movement stresses at the end joints. Expansion joint failures can be catastrophic because the joint is a clear passage to the interior of the space below. For that reason, raising the joint above the plane of the waterproofing membrane will minimize the quantity of water that can flow into the building.

Membrane defects are also common. Pinholing and cratering in the single-component, tar-modified urethane coating depicted in Figure 14.6 are major problems with liquid-applied membranes (LAMs). See Chapter 11.A., 11.D., and 11.H. for detailed discussion of LAM defects, or Chapters 9 and 10.D. for problems with other types of membranes.

Figure 14.5 The pre-applied membrane has been damaged by a brace to support the rebars. The hole has been inadequately patched with the membrane instead of the liquid-applied component of the system

Figure 14.6 Pin-holing and cratering result when the substrate on which this LAM was insufficiently prepared. Dry blisters formed by the inclusion of air and the reaction of the urethane to the moisture burst upon heating by the sun. This left a coating on the bottom of the crater that is too thin to resist hydrostatic pressure

E. Slabs-on-ground

Failure of a slab-on-ground can cause big problems, such as the destruction of an expensive hardwood floor. Slabs-on-ground waterproofing failures resemble roof failures in their most hazardous locations: at terminations of the slab and penetrations through them. The most common location for leakage through a slab-on-ground is at the joint between the slab and its peripheral basement wall. Other failure-prone locations include interior columns, pits, and pipe penetrations. Slabs-on-ground occasionally fail because they were structurally under-designed for hydrostatic uplift pressure.

In general, the solution for preventing common problems of slabs-on-ground is proper detailing at the termination or penetration. Where hydrostatic pressure exists, waterstops in cold joints is mandatory. Failures occur for the following reasons:

- waterstops were omitted
- waterstops were not spliced to ensure continuity (see Figure 14.7)
- waterstops in slabs were terminated short of foundation walls
- waterstops filled with bentonite were located too close to the top of the foundation and were pre-hydrated by water on the floor

Figure 14.7 Improper installation of waterstops in these cold joints in the pressure slab caused water to infiltrate to the surface. Failure to ensure continuity of the waterstops at the intersection was the problem. Construction joints create a continuous discontinuity in the slab. Waterstops are the only barrier to prevent leaking and are most vulnerable where they are spliced or joined

One common leak-prone location occurs where pits or shafts with one wall common to the foundation extend below the slab-on-ground slab. The complex geometry and sequencing of pours frequently creates discontinuities in the waterproofing membrane at changes in plane. Careful detailing requires multiple sections *and* an isometric that depicts the transitions in three dimensions.

A similar leak-prone situation occurs where tunnels or higher basements penetrate foundations with lower slabs-on-ground. The transitions between slabs-on-ground at elevations higher than the bottom of the foundation require careful study to avoid future leaks. To provide a practical, leak-free waterproofing system, consider normal construction sequence, allowing for adequate concrete curing time and for erection and stripping of formwork.

You should also minimize pipe and drain penetrations, which are frequent leak sites in waterproofed slabs-on-ground. Pipes should penetrate foundations and drainage should be accomplished via pits with horizontal pipes. Where penetrations are unavoidable – e.g., at interior columns and rock anchors – indicate waterproofing in great detail.

F. Preventative strategies and remedies for below-grade waterproofing failures

Issue: Sheet waterproofing systems have faulty seams.
Remedy: Ensure that seams of self-adhesive sheets are rolled. Ensure that seams on HDPE are taped.

Issue: Unsupported sheets under pressure slabs and in blindside waterproofing on walls allow seam failure.
Remedy: Specify that walls of depression be sloped at 45 degrees, not cut square. Mechanically compact substrate.

Issue: Horizontal cold joints, such as those from interrupted pours, are not provided with waterstops.
Remedy: Note in concrete specs and shop drawings that waterstops be installed in *all* joints, not just "where indicated."

Issue: Failure to remove soil and debris from HDPE membranes under pressure slabs.
Remedy: Specify that a three to four-inch thick concrete protection slab be cast over the HDPE on the same day that seams are completed.

Issue: Well points or similar pipe penetrations through slab located too close to foundation to properly flash.
Remedy: Require dewatering shop drawings and locate all well points as well as piping.

Issue: Blockouts in foundation walls for beams reduce wall thickness so much that it is insufficient to resist water or backfill pressure.
Remedy: Increase wall thickness at beam pockets.

Issue: Backfilling insufficiently compacted or in lifts that are too high, resulting in shear failure of membrane.
Remedy: Insist that backfilling be reviewed with the backfilling contractor at pre-waterproofing conference.

Issue: Walers encroach on space between outboard face of foundation and earth retention system.
Remedy: Specify flowable fill. Provide holes in waler's webs for drainage.

Issue: Sheet membranes installed parallel to the footing where seams are exposed to shear failure from backfill compaction and settlement.
Remedies: Specify that sheets are to be hung vertically in about eight-foot lengths. On blind side waterproofing specify that form spreaders be installed with loss of or lack of confinement of bentonite sheets and panels due to:

- non-uniformly flat installation
- wood lagging failures including inadequate bracing and decay of untreated or inadequately treated wood
- tie-backs failure

Issue: Loss of or lack of confinement of bentonite sheets and panels.
Remedies:

- Ensure that wood lagging is treated for continuous immersion.
- Specify that panels and sheets be installed flat and unwrinkled.
- Ensure that backfill will provide continuous restraint when tie-backs are burned off.

Issue: Wide caulked vertical cold joints provide insufficient support to resist hydrostatic pressure
Remedy: Specify that joints be cut out or ground out and patched with concrete, not caulking.

Issue: Under-slab waterproofing is damaged by pipe supports.
Remedy: Specify that piping be suspended from rebar cages.

Issue: Rebars are cut with torches allowing hot ends to fall on under-slab waterproofing membrane.
Remedy: Specify that all rebars be cut with saws.

Issue: Failure to reinforce tie-ins at vertical and horizontal plane changes.
Remedy: Indicate cold joints on the structural drawings and indicate or note that they be covered with additional strips of waterproofing.

Issue: Form spreaders on blindside waterproofing are vibrated from concrete pour and lose their seal.
Remedy: Use screws, not nails.

Issue: Waterproofing is not continued above grade and top is left unprotected.
Remedy: Detail on drawings how top of waterproofing is terminated.

Issue: HDPE under-slab waterproofing is turned up at walls resulting in tenting and open seams.
Remedy: Indicate wood cants on drawings and shop drawings to provide solid backing.

G. Leak detection

Confronted with a waterproofing failure, the first step is to determine the source of the leak water. The second step is to determine the leak location, or locations, via correlations with design and events. Anticipating future leakage is much more difficult with positive-side waterproofing, invisible from the occupied space, than with negative-side waterproofing, which is visible from the interior. Though stains or blistering and peeling paint may presage leakage in positive-side waterproofing, failure is normally discovered only after leakage occurs. Negative-side waterproofing, on the other hand, often gives warning of impending leakage when cracks form in the cementitious/crystalline coatings.

The investigation starts with a review of the leak history. If the leaking intensifies soon after rainfall, that fact points to surface runoff as the leak water source. On virtually all structural plaza (or earth-covered) slabs, the leak water originates with surface runoff. Common locations for such leaks are dynamic construction joints between foundation walls and structural slabs. These joints sometimes lack the reinforcement required for extra membrane protection. Rising walls and conduits are also common leak sites. Conduits are the more common of the two, and, fortunately, easier to repair.

Constantly flowing leak water uncorrelated with rainfall indicates a more difficult problem, with a greater variety of possible sources than rainfall-correlated leakage. Leak water bubbling up between the foundation wall and basement slab-on-ground can result from a combination of surface water and a rising groundwater level. If the source is a change in direction of an underground stream or discharge from a downspout or ruptured catch basin, sealing the joint may merely shift the leak location to a weaker spot in the wall or slab.

H. Investigators' checklist

Continuous leakage may result from a leaking sanitary sewer line. To verify this hypothesis, check infiltrating leak water for its coliform count. Where the sanitary sewer also carries storm-water runoff, flash storms can cause overflow that raises the water table. Again, check for coliform count to detect the presence of sewage.

Continuous leakage can also result from ruptured water supply systems. To verify this hypothesis, test infiltrating water for the presence of chlorine.

At densely developed sites, where most of the ground is paved, surface runoff is seldom a significant factor in waterproofing leaks. The only exceptions to this general rule are cultivated areas and other unusual modifications of the typically paved urban landscape.

Leakage resulting from storm-water overflow alone, with no contribution from sanitary sewage, can indicate a larger variety of possible sources, notably:

- sheet metal roof leaders improperly connected to cast iron or tile piping systems

- broken connections to gutters and leaders, causing concentrated overflow at foundations
- clogged drainage systems (from root intrusion into subsurface drains, infiltration of fines clogging the percolating aggregate stratum, or clogged areaway drains)
- flashing failures at conduits in pools or planters
- expansion-joint rupture in plazas and connections of pedestrian tunnels to foundations
- failure of inadequately compacted cement cants, under increased hydrostatic pressure
- disintegration of bituminous, butyl, and PVC membranes from leaking underground oil storage tanks

You may also encounter more complicated causes for continuous leakage. An adjacent excavation can initiate a chain of leak-promoting events. If the adjacent well points or excavation goes deeper than the leaking space, it can lower the water table, causing differential foundation settlement. That can induce leak-admitting tensile cracks in the foundation concrete.

Defective plaza drains are another source of excess water eventually leaking into an occupied space. Drains may be negligently designed, lacking the two-level drainage feature required for closed-joint plaza construction. If plaza surface gratings are solidly connected to the drains, with no provision for differential movement, the connection can rupture in a shearing failure, allowing excess water to overflow and eventually find its way into the occupied space.

I. Prevention guidelines

Clients need to invest in quality assurance (QA) to protect their investments. A QA inspection is the only way to assure that installation has been accomplished exactly as specified. The alternative is blindly trusting verbal guarantees, or relying on manufacturers' limited warranties or litigation as a post-failure means of rectifying construction errors. You can prevent some of the most common and obvious causes of waterproofing failure – e.g., flouting of backfill requirements by the site subcontractor – merely by putting a qualified, full-time inspector on the site.

Inspection, preferably by an independent agency financed by the client, should be formalized. Dorothy Lawrence suggests the following submissions:

- daily reports with photos showing progress of the work
- location and scope of work, dated and noted on drawings
- copies of daily weather reports
- material inventory (to assure installation at the specified rate)

This documentation serves two purposes:

1. It reduces the probability of negligent installation practices, which can culminate in a failed project.
2. It gives you and your client powerful ammunition against the contractor if a failed project winds up in litigation.

When the inspector has recorded violations of the specifications along with warnings to the contractor, these records can be decisive in the courts or at arbitration hearings, where written documentation usually significantly outweighs oral testimony.

15 Remediating basement leaks

A. Introduction

Basements subject to hydrostatic pressure leak due to movement that opens cracks and joints and an unanticipated increase in hydrostatic pressure, or the failure or absence of:

- a positive-side waterproofing system;
- waterstops; or
- flashing.

The protocol for remediating these leaks includes:

- investigating their history;
- investigating phenomena that may have changed the environment;
- locating the source of water that caused the pressure increase; and
- selecting the most appropriate methods for remediation.

Solutions include:

- managing water discharged from roof drainage systems;
- modifications to site topography;
- re-waterproofing;
- installing interior perimeter drains; and
- injecting grouts into the soil surrounding foundations behind walls, under slabs, and in cracks and joints.

Leakage can make an underground space uninhabitable, and instituting appropriate corrective measures can be difficult and costly. The process can be disruptive for building occupants, damaging to the existing waterproofing itself, and it may not always be productive or cost-effective.

B. Investigation methodology

The late Jack Fielding of the British Research Establishment likened the investigation of building defects to a detective story. The building is analogous to the victim; the defect is analogous to the crime; and materials, installation, and environmental conditions are among a long list of suspects. To solve a case, a building forensic professional should conduct an investigation by taking the following procedures:

- Accumulate historical and other pertinent data, such as a geotechnical report.
- Ascertain, if possible, the composition and thickness of the basement slab and foundation, which can affect injection techniques.
- Determine whether leak sites are active or passive.
- Determine whether leaks are intermittent or continuous.
- Verify whether water is visible and flowing or the leak sites just evidence damp spots.
- Survey floor slabs for staining and walls for efflorescence.
- Survey walls for flood level lines.
- Inspect cracks for width variations, patterns, point of origin, etc.

C. Determining the cause

Leaking through below-grade concrete structural components of a basement usually shows at cracks, construction joints, and penetrations. On masonry foundation walls, it can extend over many mortar joints.

When the leaking occurs after a basement has remained dry for a long period, it is usually caused by failure of a structural element or one of the waterproofing components. These failures can be exploited by an unanticipated increase in hydrostatic pressure. Some causes are:

- omission of a waterstop, or its failure at a construction joint
- gutters or leaders are too small, allowing storm water to overflow and spill down into the soil surrounding the foundation, thus increasing the hydrostatic pressure
- absence of gutters (see Figure 15.1)
- broken roof gutter/downspout or downspout/sewer connections
- downspouts (leaders) discharging into the soil adjacent to the foundation (see Figure 15.2)
- downspouts too widely spaced or absent at eave plane changes
- a discontinuous concrete pour creating an unsealed horizontal cold joint
- membrane failure from a ruptured oil tank or petroleum based soil poisons
- tie-rod or form spreader displacement or corrosion

Figure 15.1 The absence of gutters on this house on Block Island, Rhode Island caused windows to leak and the basement to flood

Figure 15.2 All these downspouts effectively drained the gutters, but the water discharged at their bottom terminus flowed into the basement. Downspouts should terminate in a drain connected to a storm sewer or dry well or at least 12 feet away from the foundation wall

- differential movement between suspended slab and foundation
- installation of new, unsealed piping or conduits through an existing foundation after construction
- an incomplete cure of liquid-applied membranes on plazas
- plaza drain flashing failure, often caused by differential movement between the structural slab and the wearing course (see Figure 15.3)
- negative slope adjacent to the foundation
- water run-off from recently graded adjacent property
- clogged drains in areaways (light wells)
- footing drains clogged
- poor compaction of cement cants at slab/wall reentrant angle
- sidewalk curb displacement that ruptures the underslab/foundation waterproofing juncture
- nails in blindside forms backing off and penetrating the membrane
- unsupported membrane on foundations over deeply grooved control joints
- un-flashed conduits and sprinkler pipes in planters
- root penetration in planters
- unsealed end laps of metal flashing on below-grade brick shelves

Figure 15.3 Flashing in this plaza drain has failed. Salts from the quarry tile setting bed have been deposited on the drain tail piece and elbow. The drain was installed in a sleeve and the under-deck clamps omitted

Leak sources can often be identified by observing the color or odor of the infiltrating water or from chemical tests. For example, the presence of chlorine in the infiltrating water often indicatives a failure in an underground domestic water line. As a rule, waterline breaks are usually quickly repaired because the utility is alerted by complaints from users that their water pressure has dropped.

A high bacteria level (e.g. E. coli) is usually associated with a sewer line break. In contrast to waterline breaks, sewer line breaks often go unreported, because the leakage is too small to noticeably affect the discharge of effluent into the sewage treatment plant. They are often not repaired quickly because that benefits the utility. The lower volume of incoming effluent reduces the quantity of wastewater that requires treatment.

Oily stains may indicate either a ruptured below-grade fuel tank or solvation of bitumen in the waterproofing membrane (see Figure 15.4).

D. Remediation methods

The location and severity of leakage generally should determine the most appropriate remediation. Critical factors for selection are:

Figure 15.4 The oily stain is an indication that the presence of petroleum in the backfill dissolved the bituminous waterproofing system and destroyed the flashing at the footing toe

- access to the basement foundation wall and slab from the interior or exterior;
- the degree of disruption to the occupants; and
- budget constraints.

As a rule, begin with the simplest and least expensive method and progress to more complicated and costly remedies.

1. Above-ground building components and topography

Remediating and reconfiguring roof edge drainage systems and the topography of the land adjoining a building can reduce hydrostatic pressure. It is an inexpensive method to stop or reduce infiltrating water because the building envelope and land are readily accessible.

Consider some or all of the following steps before attempting more complicated and expensive methods. To stop water infiltration, for example:

- install roof gutters where they are absent.
- install or augment the size of downspouts (leaders). Connect downspouts to a storm water sewer or extend to daylight at least 10 feet away from the building.
- re-contour the grade to provide positive drainage away from the building – generally one to two percent for at least 10 feet.
- construct swales to divert water around the building.
- install an interceptor drain uphill from the basement.
- install a perforated drain pipe in a gravel bed 12 inches long at the building perimeter or on a continuous concrete pad at the bottom of metal and glass curtain walls. This reduces hydrostatic pressure or diverts water away from the foundation.
- reposition sprinklers that spray water on walls above the cap flashing.

2. Draining water

Strategies for draining water within a basement include installing interior gutters within or under the perimeter of the slab to channel the infiltrating water to a sump from where it can be pumped to a sewer. Where basement walls are shallow, perforated pipes should be installed outboard of the foundation at the footing and the infiltrating water pumped to sewers.

The "draining water strategy" is appropriate for residences using unskilled labor, but care must be taken to ensure that the interior gutter is watertight and sloped sufficiently to provide positive drainage to the sump.

Inaccessibility is a disadvantage of the draining water strategy. If the gutter is behind furred finishes, standing water and elevated humidity can breed mold, cause unpleasant odors, and corrode fasteners. As few commercial or industrial below-grade spaces are free of intersecting walls or applied finishes, walls may be inaccessible due to the presence of wall-mounted piping or the mechanical equipment's close proximity to the wall. That makes a perimeter gutter impractical.

A second drainage method is to install drainage composites on the wall and floor slab and cover them. The drainage composites are covered with watertight panels and the floor with a three- to four-inch concrete slab. The water contained within the drainage composite is conducted to a sump and removed by a pump to a sewer.

3. Repairing or replacing positive-side waterproofing (also see Chapter 7)

Repairing or reinstalling waterproofing on a concrete foundation can be costly, but is effective on shallow foundations and where the foundation is easily accessible. If foundations can be exposed on the outboard face, they should be parged or coated to produce a smooth uniform surface, and then waterproofed. Parging is also a prerequisite on the inboard face where negative-side waterproofing is to be applied.

The flashing on brick shelves below grade is often inadequate. Either it is not carried high enough above grade or the end laps are unsealed. A straight line of dampness on the interior of the foundation wall is often indicative of this condition. Replacement of the flashing is the safest corrective measure.

4. Crystalline and cementitious materials

Negative-side waterproofing materials; such as crystalline coatings, oxidized metallic, breathable epoxies and micro and hydraulic cements; can be applied over the interior face of foundations and slabs on grade.

Coatings are efficacious where water is infiltrating through many locations over a large area and injecting grout curtains behind foundations or under slabs would be too difficult or expensive.

When coatings are to be applied, surfaces must be mechanically abraded to open the pores. Cracks and joints must be routed out and the crystalline grout forced into them.

Coatings and grouting are relatively inexpensive techniques that are easy to apply and quite suitable for active leaks. However, they are not recommended for finished basements because of their lack of water vapor control.

Crystalline, cementitious coatings, and some epoxies are vapor-permeable and should not be used in spaces where high humidity is unacceptable.

E. Injecting chemicals

1. Remedial bentonite injection grout

As a remedial injection grout, bentonite is less expensive than some chemical-grout materials. Injecting bentonite grout without having to excavate the foundation perimeter makes it easy to create a positive-side water barrier in an existing structure. Once injected, the bentonite grout forms into a solid, monolithic material coating the exterior of the foundation. It is suitable for masonry and concrete foundations where water penetrates through many small cracks or appears simply as dampness on the interior surface of the foundation wall.

Bentonite grout sets as a thick membrane with good cohesive gel strength; its adhesive properties secure the material in place. The grout remains pliable and will not rupture with subsequent concrete cracking. The grout has no exothermal reaction that can degrade the material, making it suitable for filling large-void areas adjacent to the structure.

High-solids bentonite grout is pumped in a fluid state in contact with the exterior of an existing structure where it sets to a solid material, forming a water barrier layer around the structure. Grouting is accomplished by one of these two methods:

1. Injecting grout through a tremie pipe that has been manually worked down into the soils at the exterior face of the foundation wall from grade (as opposed to excavating).
2. Injecting grout from the interior of the structure through a pattern of drilled holes through the wall.

The exterior tremie pipe injection method is typically employed on shallow foundations of one story below-ground – 10 feet (3m) or less. The interior injection method is used with deep foundations and when injecting under a slab on ground.

2. Injecting sodium bentonite into the soil from the interior

Sodium bentonite clay can be hydrated to form a grout slurry that is injected into the soil adjacent to the foundation. When the moisture induced expansion of the clay is restrained, the hydrated gel becomes dense and watertight.

Sodium bentonite grout can be injected behind foundation walls from the basement interior or from the exterior of the building without excavating. It can also be injected under basement slabs. It is particularly effective for grouting between the bottom of foundations and the top of underpinning (see Figure 15.5). Injecting from the interior requires pre-drilling holes through the concrete wall or slab to pump the grout through. Exterior injection is accomplished without excavating by manually sinking a pipe down into the soil adjacent to the foundation. The injection pipes are spaced two-feet on center and generally used for shallow foundations (less than 10 feet deep).

Pressurization is critical with sodium bentonite grout. Often the top foot of soil is inadequate for confinement for the grout to seal as the sodium

Figure 15.5 This existing basement was lowered and the walls and floor slab waterproofed with crystalline waterproofing. The joint between the underpinning and the existing foundation had been dry-packed. The joint as shown above is being injected with bentonite grout

bentonite can escape during placement through cracks in the soil. When injected from the exterior, it can escape through the annular space between the pipe and the soil. Without confinement, the sodium bentonite cannot create a sufficient pressure. When using sodium bentonite to seal stone foundations, air pockets in dry laid up stone or mortar can prevent the grout from achieving complete penetration.

Sodium bentonite grout is less expensive than urethane or acrylate esters and requires less application skill. It can fill large voids in the soil, and is suitable for remediating failed waterproofing, especially bentonite. It may be used for sealing leaks in thick stone foundations without excavating, but may not fill small voids or thin cracks between stones as completely as chemical grouts do. It is particularly effective for grouting between the bottom of foundations and the top of underpinning.

Advantages of sodium bentonite grout are that it:

- is a natural, inert mineral and contains no VOCs
- is placed at low pressures
- can fill both small and large void areas around the foundation
- remains pliable enough to shift with the structure without rupturing, thus maintaining its watertight barrier

- can be injected through the same drilled holes to provide greater coverage if the initial injection fails to stop water infiltration
- does not stain concrete, and cleaning equipment and the work space is easy

Disadvantages of sodium bentonite are:

- Unregulated high injection pressure can damage un-reinforced concrete or lightly reinforced masonry walls.
- Injections can be washed away by flowing water in well-drained gravel.
- It may also flow into footing drains and clog them.
- When injection fails to stop the infiltration, additional sodium bentonite is the only option since it forms a bond breaker with alternate materials.

3. Injecting epoxy into joints and cracks

Epoxies are the preferred materials for repairing structural damage but are not as effective as urethanes for stopping leaks.

Epoxy should only be injected by skilled applicators who can select the most appropriate products with the correct viscosity and employ the proper injection pressure and timing.

Epoxies are most effective when cracks are not subject to cyclical stresses such as temperature swings or vibration. Under these conditions, brittle concrete frequently cracks in locations adjacent to an injected crack.

4. Injecting chemical gels and foams

Urethane gels and foams are hydrophobic and hydrophilic chemicals consisting of one- and two-component resins. They can be formulated to be flexible or rigid and/or have open or closed cells depending on the selection of resins and their proportions.

Acrylate polymers are two-component chemicals that cure without forming a gel or foam phase that would change their density. These chemical gels and foams function as a flexible gasket between surfaces, although hydrophobic urethanes do not adhere to the surfaces being repaired as well as hydrophilic polymers do.

A major advantage of injection systems is that they are applied from the interior side of the basement. That can stop leaking, thus avoiding the difficulties and expense of removing the surrounding soil or floor slab. They are particularly appropriate for sealing pipe penetrations and construction joints. However, they should not be used in concrete masonry because their expansive qualities can damage low density units.

The material stays flexible enough to accommodate freeze/thaw and wet-dry cyclic movement. Chemical grouts can stop leaks and permit limited joint movement, but in contrast to epoxies, will not restore strength or structural integrity.

Chemical grouts are injected into cracks and joints where they expand on exposure to water and form a flexible gasket to stop leaks. They also expand in contact with water when injected behind foundations and under slabs to form water impenetrable gel curtains. The preferred chemicals for injecting between the soil and concrete are acrylates due to their low viscosity and pumping pressures.

5. *Urethanes*

The most common urethanes are hydrophilic grout for filling cracks and joints and hydrophobic foams for filling voids in the soil behind concrete foundations and slabs. Hydrophobic urethane foams react with small amounts of moisture in the soil and expand, forming a water-impermeable grout curtain.

Hydrophilic foams are less susceptible to dilution when injected into the soil than hydrophilic foams. Urethanes in their cured state do not shrink or swell from moisture and are less susceptible to dilution when injected into cracks. Hydrophobic urethane foams react with small amounts of moisture and expand, forming a water-impermeable grout curtain. Shrinkage or swelling can occur, as the moisture content can be up to 50 percent of the cured matrix, however, adhesion of the foams in cracks allows for variations. Their cured density is less predictable than with hydrophobic foams and is affected by changing moisture content, which causes them to shrink or swell.

6. *Acrylate polymers*

An acrylate polymer is a chemical grout formed by the reaction of the A and B components. The cured grout has limited adhesion to concrete as it expands and creates a barrier. The grout will dry and shrink, but will re-expand on contact with moisture. However, since the reaction can be delayed by up to five hours to reach its fully expanded state, it is not as effective in very dry climates. Therefore, give careful consideration to using it above grade.

Properly mixed acrylates are non-corrosive and non-toxic, but are more expensive than urethanes because they do not expand as much. They can form superior grout curtains are much more effective when used close to grade since they are denser than urethanes and do not expand as much as either urethane or Bentogrout®.

Acrylate polymers are effective in cracks and joints, as well as behind foundations and under slabs. They form superior flexible grout curtains that are unaffected by moisture, which helps to compensate for their higher cost. They are very fluid and nearly as viscous as water. As acrylate polymers exert less pressure than sodium bentonite and most urethanes, they are suitable for injection into thin cracks and joints in concrete and masonry, and for repairing structural foundations and deteriorated membranes in split-slab assemblies.

7. Gel and foam installation

To create gel curtains, holes are drilled through a slab or foundation. Ports (also called packers) are installed in an offset diamond pattern, typically spaced 18- to 24-inches on center into or through the concrete slab and foundations. The chemicals are injected under pressure (see Figure 15.6).

To stop leaks at cracks and joints, the holes are drilled part way into the concrete at opposing angles, staggered and alternating on either side of the crack or joint, progressively from port to port. The chemicals are injected under pressure until they extrude from the adjoining crack or hole.

Figure 15.6 Chemical injection

One of the requirements for successful injection with hydrophilic grouts is experienced technicians that look for situations where the grouts are not adequately confined and have the skills to adjust to the situation, for example in joints that are wider than hairline cracks.

Before drilling, the contractor should ascertain whether pipes or conduits are buried in the slab or wall. See “Appendix C: Waterstops” for preinstalled hoses for injection.

An ASTM Standard Guide for Waterproofing Repair of Concrete by Chemical grout Injections is forthcoming.

16 Specifications

The proliferation of complex new materials steadily increases the challenge of specifying the most appropriate waterproofing product or system for specific projects. Traditional materials – built-up bituminous membranes, bentonite, cementitious coatings, and LAMs – are still being used. Newer choices include modified bitumens, synthetic rubber and plastic sheets, polyester liquid-applied formulations, and laminated bentonite sheets, among others.

Choosing the most appropriate method of waterproofing is critical due to the extremely high expense caused by below-grade waterproofing failure. Removal and replacement of a waterproofing system often also requires removal of tons of and backfill. If access issues prevent reconstruction, the alternative is negative-side waterproofing injections with hydrophilic materials. This repair method is also very expensive, and not always satisfactory. For these reasons, the initial cost must almost always be your least important selection criterion.

Perhaps the most common mistake in waterproofing specifications is verbatim copying of a manufacturer's specification. Every waterproofing project has unique aspects that require customized specifications. Copying a manufacturer's specification has two additional specific drawbacks:

1. It may eliminate competition by introducing unique, proprietary requirements.
2. It complicates processing requests for acceptable substitutions.

A. Division of responsibility

The typical waterproofing project involves at least four principals – architect/engineer (A/E), manufacturer, general contractor, and waterproofing contractor. There is frequently a fifth, the construction manager (CM). Waterproofing requires coordination among these principals because it is almost always installed in stages. Scheduling these stages and assuring that the parties involved meet their obligations in a timely fashion are critical.

Follow these guidelines to assign responsibility:

1. **The A/E** bears ultimate responsibility for selecting an appropriate waterproofing system, based on soil reports and other factors discussed in other chapters. You must:
   a. verify a system's suitability with its manufacturer;
   b. define the scope of work; and
   c. provide details for all conditions.
2. **The manufacturer of the waterproofing materials or system** bears responsibility for:
   a. providing the proper grade of materials (e.g., temperature-dependent or salt-water resistant materials when appropriate);
   b. checking soil reports to assure that its products will not be damaged by soil contaminants; and
   c. assisting in the preparation of shop drawings.
3. Bearing the greatest range of responsibilities, **the general contractor** must:
   a. schedule the installation of building components on which waterproofing is applied, the waterproofing itself, and backfill operations;
   b. verify the conformance of backfill aggregate with specified size limits and proper compaction;
   c. maintain a water-free excavation;
   d. assure that the concrete contractor follows specifications for surface preparation, concrete curing, and form-release agents;
   e. restrict traffic over slab waterproofing by plumbers, electricians, and other workers;
   f. assure the provision of temporary protection of waterproofing at pipe scaffolds or masonry installations when walls are erected above the waterproofing membrane.

   The general contractor's control of scheduling is vital in coordinating the work among the various contractors. Scheduling of waterproofing installation is dependent on the scheduling of backfill operations, as the backfill often serves as a platform for installing waterproofing on foundations.
4. **The waterproofing contractor** is responsible for:
   a. installing the work in accordance with the manufacturer's instruction and the contract documents.
   b. examining the waterproofing substrates (concrete or masonry) and verifying that their texture, soundness, and dryness are suitable.
   c. coordinating work with the concrete contractor to assure application of cementitious coatings while the concrete is still green.
5. **The CM** must exercise caution in applying the criteria for Value Engineering (i.e., achieving a dependable waterproofing system at least cost).

B. Specifying unfamiliar products

Because of the high cost of a waterproofing system's failure, you should consider new and unfamiliar products guilty until proven innocent.

By pursuing a conservative design policy, you avoid selecting an untested or a new system from a manufacturer with little experience in waterproofing. A more difficult problem arises when a well established manufacturer modifies a proven product, markets a new product under an old name, or markets one from a company it has purchased. If the manufacturer is foreign, the situation is even riskier. In any event, the original, or vendor, if not original manufacturer becomes a de facto third party to the construction contract. This complication muddies the legal waters, which are murky at best in the U.S., making it more difficult to determine responsibility in litigation.

To avoid these complications, you should require the manufacturer's assurance that the product has not been modified. At the least, you should request an exact description of any product modification, an explanation of its purpose, and comparative test results.

If for some unforeseen circumstance, you must consider a new or drastically modified product, you should investigate as follows:

1. Obtain from the manufacturer a list of projects on which the product has been previously used, including the names of the client, A/E, and waterproofing contractor. By interviewing these principals, you can ascertain whether the conditions on their projects are similar to your project's conditions.
2. Ask the manufacturer's representative about conditions in which the product is not recommended, and about the failures as well as its successes. Even if the manufacturer does not candidly inform you about the product's shortcomings or failures, you can testify accordingly during litigation over the product's failure. You are better protected if you made the attempt.
3. Notify the manufacturer in writing how you intend to use the product.
4. Request relevant technical data from the manufacturer. If it is not supplied, you have an excellent reason for not specifying the product.
5. Investigate the manufacturer's past performance as much as possible, seeking answers to questions such as:
   a. Does the manufacturer warrant the product's performance?
   b. Has the manufacturer produced previous waterproofing products?
   c. What is the company's record for manufacturing and marketing reliable waterproofing?
   d. Does the manufacturer require applicators to prove competence via licensing or approval programs?
   e. Has the applicator successfully installed similar products?

Before specifying a new product, you should inform the client (preferably in writing) about possible risks, as well as the advantages of using the new product. If the client is unwilling to accept these risks, specify a more proven material.

C. Organizing waterproofing specifications

1. Waterproofing specification standards

Specifications for waterproofing fall under Division 7 in the CSI (Construction Specifications Institute) MasterFormat. Bituminous, bentonite, plastic, rubber, cementitious and all other materials for below-grade waterproofing systems are grouped under sections numbered in the 07100 series.

Plastic sheets used for dampproofing under slabs-on-ground are often specified under Concrete – 03300, but the CSI Master Format lists them under Section 072616. Admixtures for integrally waterproofed concrete and waterstops are also specified in this Section. Insulation, drainage composites, and protection materials are included in waterproofing sections.

AIA MasterSpec and CSI Monographs contain sections on waterproofing for some systems. However, you should be aware that the technology of most current systems is constantly changing. Referring to a manufacturer's current specification, or better yet, information obtained from an interview with a technical representative is always the wisest action.

In addition to Division 7, which covers waterproofing systems per se, Divisions 1–3 are also involved in specifying waterproofing.

a. Division 1 – pre-installation meeting

Section 01200. A pre-installation meeting is as necessary for waterproofing as for roofing. At this meeting, require the following attendees: A/E; inspector; general contractor; subcontractors for waterproofing, concrete-casting, backfill, formwork, excavating; foremen for all contractors and subcontractors, plus representatives of manufacturers of materials to be incorporated into the work. If a green roof is part of the design or if trees are to be installed over waterproofing, the landscaper's presence is desirable to review methods of staking, installing root barriers, drainage, and similar related subjects.

This pre-installation meeting should be scheduled prior to excavation and cover:

- width of the foundation excavation, to allow adequate space for installation of positive-side waterproofing;
- form release and concrete curing agents;

- sequencing of concrete-casting for footing and slab-on-ground, structural slab, and backfill operation; and
- the waterproofing contractor's review of required concrete finish and surface moisture requirements.

A second meeting should be scheduled after the concrete is cast. Its purpose is to allow concrete and waterproofing contractors to inspect substrate surfaces, to assure their suitability for waterproofing application.

A third meeting may be required to re-inspect the substrates after remedial work on the substrate surfaces has been completed.

b. Division 2 – backfill

Because the satisfactory performance of a waterproofing system often depends on prompt and proper backfilling, specify the following in the appropriate Division 2 sections:

- Require backfilling as soon as possible after positive-side waterproofing is installed and covered with associated materials.
- Limit backfill lifts to a maximum height of 12 inches.
- Require compaction of backfill against bentonite panels to 85–90 percent maximum density per ASTM D1557-12E1, "Standard Test Methods for Laboratory Compaction Characteristics of Soil Using Modified Effort." Backfill over bentonite-filled tubes must be hand-placed.
- Require dewatering of areas to receive bentonite and membranes above the toe of the mud mat or the top of the footing to assure that surfaces are sufficiently dry for application of the waterproofing system.
- Specify single-graded aggregate, not less than ¾-inch size, as a granular base under slab-on-ground and porous backfill.
- Specify filter fabric and porous backfill and subsurface drains.

c. Division 3 – concrete

Because the success of a membrane system often depends on proper substrate preparation, specify the following in the appropriate Division 3 sections:

- surfaces to be waterproofed
- scheduling of concrete pours (to assure adequate curing time)
- curing materials that will not inhibit the bond of adhered membranes
- the finish (float, steel trowel) required for the selected waterproofing system. Note that some adhered waterproofing systems require extension of the membrane over the footing and down the toe. These surfaces, which are not usually finished, should be noted in the specifications

- cement cants and chamfered corners if required
- waterstops (See Appendix C.)

Also take these steps:

- Reference ASTM D5295, "Standard Guide for Preparation of Concrete Surfaces for Adhered (Bonded) Membrane Waterproofing Systems" for preparation of concrete surfaces.
- Require the concrete contractor to exercise care when casting concrete on previously waterproofed surfaces.
- Require that concrete decks to receive adhered waterproofing be cast over slotted metal decks to ensure downward drying.

When pour stops are used on slabs-on-ground, warn the contractor not to puncture the vapor retarder or the underlying waterproofing and to notify the waterproofing contractor if the waterproofing is punctured.

d. Division 7 – dampproofing and waterproofing

(1) Part 1 – General

- Reference appropriate sections in Divisions 2 and 3.
- Under submittals, require shop drawings for membranes and bentonite. Drawings should indicate all conditions, transitions, and penetrations. Include such items as number of plies, required cants, nailers, and flashing materials. Require product data submission, including installation instructions. If sheet membranes on plazas are compartmentalized, require submission of as-built drawings indicating exact locations of each divider.
- Require submission of Material Safety Data Sheets (MSDS) and stipulate that they be kept on the site.
- Require the waterproofing manufacturer to review soils reports and to state in writing that the soil chemicals will have no adverse effects on the waterproofing.
- For quality assurance on all except hot-applied built-up membranes, require the manufacturer to certify that the product or system has been manufactured and marketed in the United States for at least 10 years and that the formulation has remained consistent during that time.
- Require the manufacturer's certification of the applicator.
- Require the applicator to test for dryness of the substrate using written instructions provided by the manufacturer.
- Require an inspector retained by the client for full-time supervision of the installation.
- Include limitations on the shelf life of materials and temperature limitations for stored materials. Require flat storage of plastic and rubber rolls. Require that unboxed bituminous rolls be stored on end. All

materials – including membrane, bentonite, and negative-side waterproofing products – must be kept dry.

- Require temporary protection of waterproofing if other trades people will execute work over it or when masonry installation on waterproofed brick shelves will be delayed.
- Specify ambient temperature limitations during application. Ban waterproofing application in rain, imminent rain, high winds, or other conditions specified by the manufacturer.
- Specify sequencing and scheduling for situations such as when foundations are waterproofed in vertical stages with intermediate backfilling or when walls are erected on plaza waterproofing prior to surfacing.
- On large complicated waterproofing installations, require the manufacturer's technical representative to conduct a seminar on site consists of installing a representative application of the waterproofing assembly including reentrant angles and at least one penetration.

Warranties

- Before writing warranty requirements, read the limitations imposed by the manufacturer; especially the disclaimers. Inform clients that waterproofing warranties do not offer as much protection as roofing warranties often do.
- Most manufactures will warrant their product, but not its installation. With rare exceptions, warranties for sheet membranes exclude seams that are the most vulnerable component of a waterproofing system. Manufacturers of LAMs warrant only the product as delivered.
- Few manufacturers will include removal of overburden for examination, testing, and replacement. The primary exceptions are for waterproofing on plazas surfaced with pavers on pedestals. Overburden can range from an easily removable foot of earth or brick pavers in sand beds to four-inch-thick travertine paving in a setting bed on a three-inch-thick protection slab. Removing 10 feet of backfill against a foundation is even more costly.
- A few manufacturers offer extended warranties with the proviso that the buyer retains an independent third-party service for regular and complete inspections during all phases of installation. The third-party inspection service must be approved by the manufacturer.

(2) Part 2 – Products

- Membranes: Specify ASTM D449, Type 1 for asphalt; D450, Type I for coal tar pitch; D2178, Type IV for glass felts; and D1668 for woven glass. Use ASTM D1668, Type II, woven-glass fabric or ASTM D227 tar-saturated organic felts with coal tar pitch rather than glass felts. Cotton (ASTM D173), Burlap (ASTM D1327), and asphalt organic felt (ASTM D226) are not recommended.
- Specify ASTM C836 for liquid-applied membranes.

- Specify ASTM C578 Type VII extruded polystyrene with minimum compressive strength of 60-psi for rigid insulation on framed slabs under wearing courses. Specify Type I expanded polystyrene (0.90 pcf) as a protection layer on vertical surfaces or Type VI extruded polystyrene to insulate below-grade foundation walls.
- Require special winter-grade adhesives and primers for installation below 40°F.
- Specify thixotropic grades of liquid-applied membranes for flashing vertical surfaces or use elastomeric sheets.
- Specify accessories, including protection boards, drainage panels, termination bars, fasteners, filter cloth under concrete protection slabs, and closed-cell neoprene tubes or proprietary tubes to support membranes at expansion joints. (Prefabricated expansion joints are not advisable unless they are designed or protected from deforming under surcharge loads.).
- Specify treated wood for cast-in nailers.

Elastomeric sheets: Specify butyl sheets for loose-laid and fully adhered installations. Specify reinforced, cast PVC sheets for loose-laid or fully adhered on horizontal installations.

Beware "or equal"

See Chapter 5.D. for discussion of the dilemma of specifying materials for public projects when regulations require "or equal" wording. Regulations for taxpayer-supported projects often require that specifications list at least three manufacturers; or, if only one manufacturer is specified, you must list the physical properties of the product or system and add the phrase "or equal" or "approved equal." This can pose a dilemma.

Several membranes that are frequently specified for plazas are similar enough to satisfy this requirement. However, most membranes currently manufactured for use on below-grade foundations and pressure slabs are unique and have dissimilar properties. Moreover, their application differs from all other products. Therefore, you can neither specify *equal* products nor even list several manufacturers for a single project that can satisfy this requirement.

One solution is to write a section specifying a single product followed by separate sections for alternate products. This simplifies the specification problem but adds the dilemma of which product to depict on the waterproofing drawings.

(3) **Part 3 – Execution** (**for Surface Conditions**)

For membranes that must be applied to "dry" surfaces, specify the testing method. Acceptable moisture level should be determined by the membrane manufacturer, with testing performed by the applicator.

ASTM Standard F710, "Standard Practices for Preparing Concrete Floors to Receive Resilient Flooring" lists several test methods, including:

1. Presence of condensation under a plastic sheet 24 hours after it is taped to the slab.
2. Conductance moisture meters.
3. Laboratory test to determine relative humidity of a sample.
4. Proprietary methods that measure water vapor transmission through slab.

Use two or more methods to cross-check results and make tests in each 1,000 square feet of slab. The most tried-and-true test is to adhere a patch of the membrane and try to pull it up (see Figure 11.4).

2. Surface preparation

a. General

Require installation of flashing and reinforcing prior to membrane application.

b. Adhered systems

Remove all contaminants, dust, dirt, and laitance (fine particles) on surfaces designed for adhered systems. Reinforce corners and construction joints on substrates receiving self-adhering membranes. Remove curing agents that may limit adhesion. Sodium silicate, animal-fat-based agents, and other agents require special removal techniques. Fill honeycombs, rock pockets, indentations, and tie holes with patching cement.

On all types of adhered sheet membranes, require repair (i.e., cut-out and patching) of fishmouths and wrinkles. Self-adhering sheets and glass felts have "memory" and cannot be pressed flat.

Require priming of surfaces to receive bituminous and modified-bitumen membranes. If prefabricated, self-adhering membranes are not installed on the same day; they require re-priming.

On membranes under pressure slabs, remove mud and other contaminants from surfaces designed for waterproofing. Groundwater or heavy rains in the excavation will sometimes flood the top of footings and leave a residue of mud when the water recedes. Require water-jet removal of mud and other materials that would inhibit contact between the adhesive coated or geotextile covered membrane and concrete.

c. Bentonite

Remove mud and other contaminants from surfaces designed for bentonite waterproofing. Fill honeycombs, rock pockets, indentations, and tie holes with bentonite gel. Patch voids in lagging that are larger than ½ inch.

Require troweling of bentonite gel on joints and penetrations. Use bentonite-filled tubes at foundation wall-footing joints where panels terminate. Lap panels and sheets. Replace or repair damaged panels. Protect system from water until slabs are cast and backfill is in place.

d. Liquid-applied membranes

On concrete surfaces that will receive liquid-applied waterproofing, "detail" cracks wider than or equal to the membrane thickness by applying a strip of fabric over the crack.

Some urethane LAMs may require priming with a solvent-thinned coating of the same material.

e. Negative-side waterproofing

Acid-etch or sandblast surfaces designed for crystalline or cementitious negative-side coatings if they are applied after concrete has cured. Require dry-packing at joints. Require mechanical compaction of materials at slab/foundation joint.

3. Application

a. General

- Require immediate installation of protection boards if other workers will perform work over it or when masonry installation on waterproofed brick shelves will be delayed.
- Specify the type of finish (float, steel trowel) required for the specific waterproofing system selected.
- Where reinforcing bars penetrate the membrane – e.g., for column dowels at interior spread footings or for rock anchors – seal sheet membranes by figure-cutting the sheets around the bars and sealing them to the bars with the system's liquid or mastic component. Liquid-applied membranes are self-flashing. Do not wrap bars with membranes. They cannot be sealed to the bars' deformed surfaces.

b. Built-up membranes

- Require a visible thermometer and thermostatic controls on all kettles, set to manufacturer's recommended limits. Require rejection of bitumen heated above the specified maximum and reheating of bitumen too viscous for mopping.
- When an engineering validation test (EVT) is provided by the supplier, require hot bitumen temperature at application within ± 25°F of EVT.
- In cold weather, require double or insulated lines and insulated bitumen carriers. Store felts in a warm enclosure.

- Check for continuous, uniform interply hot moppings, with no contact between adhered felts, and average interply mopping-weight tolerance of ± 15 percent.
- Require manual brooming or squeegeeing following six feet or fewer behind unrolling felt.
- Require immediate repair of felt-laying defects: fishmouths, blisters, ridges, or splits.
- Terminate day's work with complete glaze-coated seal stripping.
- Require back-nailing on walls. Use woven-glass fabrics for flashing and corner reinforcing.

c. Liquid-applied membranes

- Specify application on structural slabs to conform to ASTM C898.
- Require two coats, with thickness and tolerances as a percentage or in mils, not in liquid content per unit area.
- Require delay in installation of protection boards on one-component, liquid-applied waterproofing until membrane has fully cured.
- Specify non-vulcanized neoprene or polyester reinforcing for flashing.

d. Self-adhering sheets

- Require applicator to roll sheets and double-roll seams after application. Require mastic on all edges at day's end. Require liquid membrane on tee joints.
- Require prompt application of protection course.
- The bottom of post applied membranes should terminate on the footing. The top of the footing and toe are usually too wet or contaminated to permit adhesion. Use the liquid membrane or mastic component of the sheet system as a fillet at the termination (see Figure 16.1).

e. Rubber and plastic sheets

- Require that applicator provide for relaxation of sheets before installation.
- Do not permit bonding adhesive on field seams. Require applicator to use a steel roller to compress sheet and seams after membrane installation.
- Require cover strips on tee joints and preferably on all joints.

4. Field quality control

a. General

- Prohibit testing if freezing temperatures can occur.
- Do not permit fire hoses or power equipment to direct stream across laps.

Figure 16.1 Liquid membrane being applied as a fillet at the membrane termination

- Make area off-limits until the membrane is protected.
- Require manufacturer to observe flood testing and to supervise repairs if needed.
- Require waterproofing contractor (or preferably inspector) to submit daily:
  (1) reports of work done with location and scope noted on a drawing;
  (2) photographs; and
  (3) weather reports.

b. Testing

These tests should be specified for waterproofing membranes:

- Test the substrate to ensure that it is sufficiently dry to accept the membrane. See Chapter 11.
- Test the thickness of a LAM to verify that it meets the specification. See Chapter 11.
- Test welds of single-ply membrane seams. See Chapter 10.
- Verify the watertight integrity of plaza waterproofing. See Chapter 13.

- Require flood testing of horizontal waterproofing on framed slabs, per ASTM D5957, "Standard Guide for Flood Testing Horizontal Waterproofing Installations." Water should be sprayed first at perimeters, to check proper flow to drains and to detect ponding.
- Alternately require electronic leak detection testing, per ASTM D7877 "Standard Guide for Electronic Methods for Detecting and Locating Leaks in Waterproofing Membranes. See Chapter 13 "Plaza waterproofing." Some leak detection systems require a grounding screen under the membrane. Specify that the screens are to be included as part of the cost of the testing system.

Specify damming method, if other than prescribed in ASTM D5957. Dams can be constructed from sand or water-filled fire hoses, planks assembled in an "L" shape, or foamed insulation boards covered with PVC, EPDM, or polyethylene sealed to the membrane and weighed with sand bags. Require cushioning of dams to prevent damage to membrane from sharp edges. This is important where a sheet carried over the top of a wood barrier is subjected to an increasing combination of shear and tensile stress from the wood barrier's sharp edge as the weight of flood-test water increases.

Require continuous supervision below the test area until it is drained. Flood-test water should generally be ponded to a maximum of four inches deep over horizontal membranes and held for 48 hours. After the water has been drained, inspect all seams of sheet-applied panels to check for water entrapment in these seams. Test cuts or moisture-meter probes can also be used if the membrane's watertight integrity is suspect. Preferably drain water into internal drains, not over or into finished surfaces.

c. Liquid-applied membranes

- Require test cuts and wet-mil thickness check for each square (100 square feet).
- Require the contractor to maintain an inventory to assure that materials are installed at the specified rate.

d. Single-ply membranes

- Require test cuts across the seams.
- Require seam probes.

17 Detailing waterproofing

One definition of "detail" is "treatment of a subject in individual or minute parts." Detailing waterproofing is your responsibility. You must provide sufficient detail to contractors to prevent your considerable risk of liability when waterproofing fails.

In detailing waterproofing, your first decision is whether to show waterproofing on the structural drawings, the architectural drawings, or both. The third option is the worst. It maximizes the probability of conflicts, as every modification made on one set of drawings must be copied to the other. I recommend that the architectural drawings show the extent of waterproofing plus details, with a reference note on the structural drawings. For a large complicated project, prepare a set of waterproofing drawings that includes plans, small-scale and large-scale sections, and penetration details.

The architectural or waterproofing drawings should also show the mud mat, drainage medium, vapor retarder, footing drains, insulation, and tie-ins to the ground floor exterior wall and plaza wearing surface, if applicable. The structural drawings should show cementitious cants or fillets at reentrant angles and chamfered external corners, and other special conditions associated with the concrete work. Failure to include these items on the structural drawings can spawn controversies on the site and possible extra costs.

Of greater importance is for the plumbing or site drawings to show the footing drains in the plan (including drainage from leaders) and connections to site utilities. The omission of this critical information can create jurisdictional disputes, and can result in omission of its directives from the project. You may bear the blame.

Clearly indicate the extent of the waterproofing. Specifications that merely require waterproofing of foundations, or drawing notes indicating "waterproofing," with an arrow pointing to the foundation cause misinterpretation and unnecessary costs. Does the membrane stop on the footing, continue across it, or extend down the toe? Does it stop at the top of the foundation, return over the brick shelf or under the coping? Is negative-side waterproofing carried into blockouts for intermediate slabs? Where does it terminate at intersecting shear walls? Does it cover the top of equipment curbs?

To prevent such questions from being answered in the field, indicate the extent of waterproofing on small-scale sections on all subsurface building components. Use a heavy dashed line to avoid confusing the waterproofing with the outline of the waterproofed component: foundation wall, footing, slab-on-ground, etc.

Draw sections at a minimum scale of 3/4 inch = 1 foot, details at a minimum of 1 1/2 inch = 1 foot, and preferably 3 inches = 1 foot at penetrations, drains, dynamic joints, and changes in plane (e.g., internal and external corners). For membrane waterproofing, indicate the arrangement of plies and reinforcing on large-scale drawings. In general, all components should be drawn, with one notable exception. You can simplify the drawing by noting, rather than drawing, protection boards wherever a membrane is shown.

Dampproofing and vapor retarders should be distinguished from waterproofing and one another, if they all exist on the same building. Variations in dashed lines such as long and short dashes, dash-dot, etc., will avoid confusion when, for example, a foundation waterproofing membrane and under-slab vapor retarder are combined in the same detail.

The applicator, in concert with the manufacturer, should submit shop drawings to ensure that all details are included, but this does not relieve you of the responsibility to provide detail drawings. Beware of applicators that simply submit cuts from the manufacturer's catalogue without customizing them to suit unique project conditions. Assure that all conditions are indicated, including:

- critical details of plane transitions and penetrations;
- membrane transitions between pressure slab and foundation walls;
- transitions between under-slab membrane and pits;
- terminations at the top of foundations, caissons, and grade beams; and
- transitions between pre-applied and post-applied membranes.

One of the most critical details is the flashing of a plaza waterproofing system at window walls that finish flush, or almost flush, with the wearing surface. Other critical, and often overlooked details, are sealing of conduits and piping in fountains and planters, and the transition of expansion joints at plane changes.

With the exception of top-down basement construction, components of the soil retaining system (SOE) require bracing to resist earth and water pressures. Generally, these are temporary and are removed during construction of the basement walls.

The components of the bracing systems frequently penetrate the membrane and require special details to maintain its watertight integrity. Generally, those include:

- tie-backs or soil nails
- rakers, walers
- cross-lot blocking

A. Sequencing

You are not responsible for sequencing. Sequencing skates close to responsibility for means and methods, a province expressly off limits to architects and engineers in their client/professional agreements. However, you must recognize the normal sequencing of construction. Foundation walls may be cast before or after slab-on-ground. Wall-bearing structural slabs, however, can be cast only after their supporting walls are cast. Waterproofing is applied to individual components and frequently follows this sequence. Constructing underground building components in a suitable sequence can greatly improve the chances for a successful waterproofing installation.

Sequencing is also important where plaza waterproofing is installed before adjacent rising walls or exterior finish materials. Brick on shelves, door sills and parapet cladding may be installed long after the membrane. Wearing courses may be applied after cladding. Wall flashing details should provide a continuous, watertight transition between counterflashing and membrane base flashing. The vertical transition of the membrane should be installed and terminated on a stub wall (see Figure 17.1).

In this example, the mud slab is extended at least 12 inches beyond the foundation or toe of a footing to permit tie-in with the vertical waterproofing membrane. However, where adhesive coated membranes and those faced with geotextiles need bonding by interfacing with the concrete cast against them, they are turned up the formwork at the edge of the pressure slab. The plane transition and the tie-in to the wall waterproofing system must be carefully detailed since the membrane on the wall (yet to be cast) must lap the membrane adhered to the edge of the pressure slab. Moreover, stiff plastic sheets – particularly in cold temperatures – may require cants to avoid a coved transition where seams can open under the weight of the fresh concrete cast against them.

When adhesive coated sheets are used for the waterproofing membrane, the adhesive is exposed to foot traffic and ponding that accumulates dirt. To protect the exposed face of the membrane, it can be covered with a three-inch protection slab prior to casting the pressure slab. The protection slab is also recommended for membranes that are faced with geotextiles whose fibers are intended to engage the concrete pressure slab cast over it.

Where the foundation wall and interior columns are designed to be supported on footings, caissons, or grade beams, the preferred detail is to extend the membrane over the mud mat and the top under the wall or column. However, the structural engineer must be consulted prior to employing this simplified approach. Some engineers refuse to permit it because the membrane inhibits the transfer of shear. Others will agree to

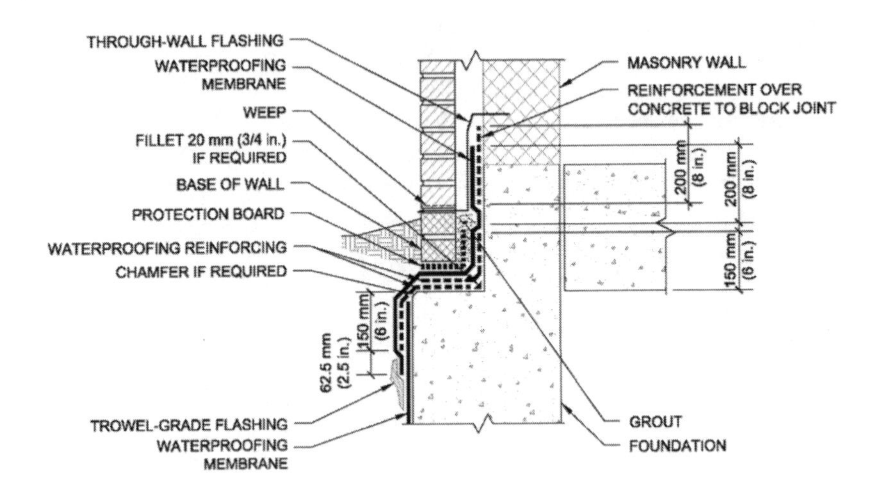

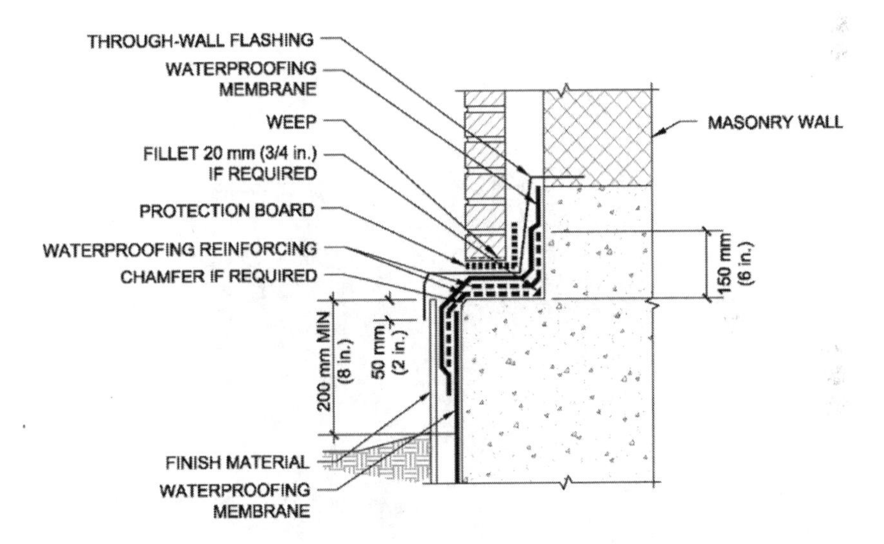

Figure 17.1 Top detail shows wall waterproofing terminating below grade. It should be extended up to the top of the stem wall and terminated at least eight inches above grade. Bottom detail shows wall waterproofing terminating above grade. It should be extended to the top of the stem wall. (ASTM International)

apply a watertight epoxy throughout the joint with two conditions: (1) the membrane is extended an inch or two into the joint and (2) the rebars are painted with the liquid component of the membrane system. When the membrane is extended through the joint, it is finger cut around the reinforcing and sealed to it with the liquid or mastic component. Note that the liquid membrane should not extend up the rebars because the deformations on the bars will prevent bonding.

Extending the membrane under the foundation over the top of caissons and grade beams avoids the laborious and torturous installation of the membrane to encapsulate the components below it (see Figure 17.2a).

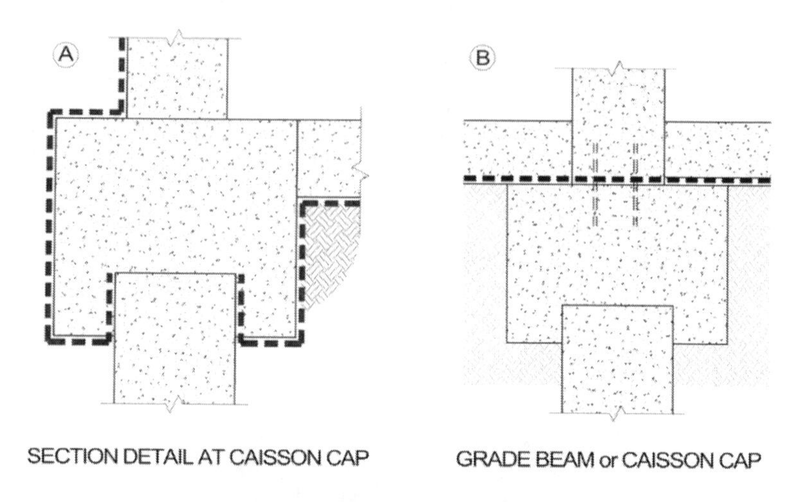

Figure 17.2 Caissons, grade beams and pile caps can be enclosed with the membrane as shown in A or extended over them as shown in B and in photo C. The latter must be approved by the structural engineer who may be concerned about transferring shear at the interface with the pressure slab. In that case the membrane may be terminated two to three inches beyond the edge of the cap. The remaining area of the interface should be coated with a waterproof epoxy, or roughened and coated with crystalline waterproofing. Rebars require sealing with the liquid component of the membrane system

The water-tightness of the joint can be maintained with waterstops and injection hoses installed at the top of the flashing and in joints as indicated.

Temporary protection of the membrane that extends under the foundation can be obtained with a protection board that is stripped before the wall is cast and left in place on the extended mud mat until the wall waterproofing is applied. The protection board will prevent damage to the membrane when the wall formwork is erected.

B. Reinforcement of membrane systems

Most systems require reinforcing at transitions – e.g., internal and external corners, and concrete construction joints (see Figure 17.3). Reinforcement is also required where reinforcing bars, pipes, and conduits penetrate the membrane. Reinforcement is usually installed before the membrane, but in some instances may be applied over it later. Refer to the membrane manufacturer's details.

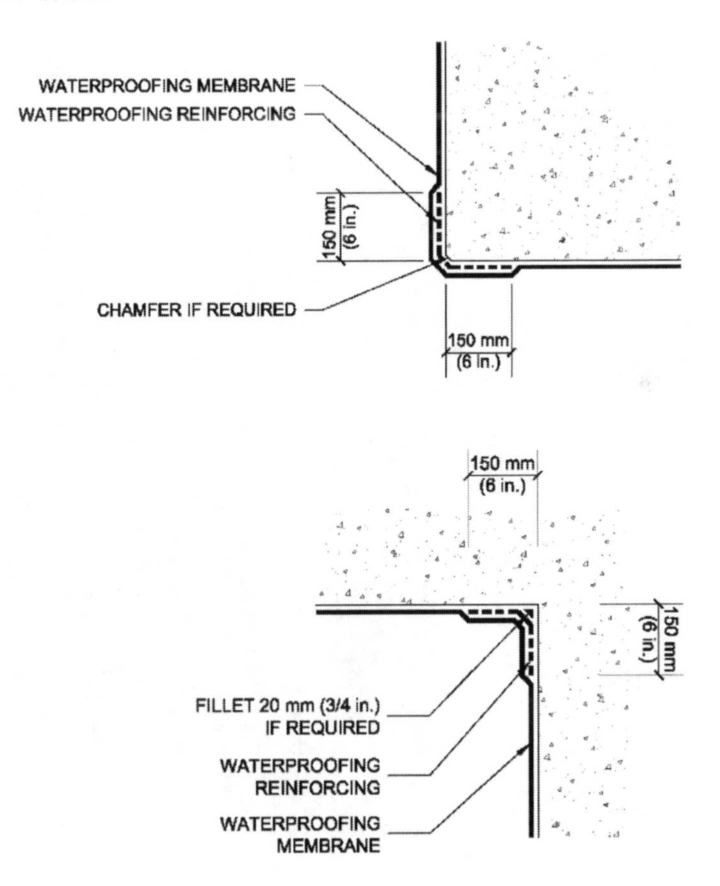

Figure 17.3 Reinforcing details are shown for exterior and interior corners

Multi-ply membranes usually require multiple reinforcement plies, whereas single-ply systems generally require only one. Additional plies are required where substrate joints may be dynamic (i.e., subject to relative movement between adjacent structural components).

When additional plies are used as reinforcement, they should extend at least six inches beyond the corner or penetration and each succeeding reinforcement ply should extend at least three inches beyond the previous ply.

Reinforcement, in the form of tape or strips applied to one or both faces of the seams, is mandatory for adhered seams (not heat-fused seams). This is where the membrane system is weakest and most liable to open under pressure from wet concrete or settlement. Seams most likely to fail are at the transition between the under-slab membrane and the foundation (see Figure 17.4).

Figure 17.4 Taping joints reinforces the most vulnerable part of the sheet membrane system

C. Termination of positive-side waterproofing

The waterproofing membrane on foundation walls that terminates at footings should be sealed at the reentrant angle with mastic or a liquid-applied sealant furnished by the membrane manufacturer. Extending the waterproofing over the footing and down to the footing bottom is not necessary. Moreover, as the footing face is normally irregular, dirty, wet, and probably partially under water, the possibility of obtaining a bond is remote.

Bentonite panels require bentonite-filled tubes at the footing/foundation reentrant angle. When the bottom of the slab-on-ground is eight inches or more above the top of the footing, the membrane is terminated at the bottom of the foundation wall. When the slab rests on the footing, the membrane joint is reinforced and the membrane is extended to the toe of the footing and sealed.

Positive-side membrane waterproofing of all types – built-up, self-adhering single-ply, and liquid-applied – should terminate at least eight inches above grade. Many waterproofing manufacturers show membranes terminating in a reglet below grade with the top edge sealed with mastic, or secured with a termination bar. *These details are contrary to good practice and irresponsible marketing by the product manufacturer.* Even when the membrane is turned into a reglet or a cap flashing is installed in a reglet, water can migrate behind the membrane through cracks in the wall above the reglet.

Waterproofing should be carried over a brick shelf or continued up the wall as indicated in Figure 17.1. Above grade, the waterproofing membrane must be protected from damage and UV radiation by cement parging, parge-covered insulation boards, noncorrosive metal, or masonry. When the facade is extended below grade, it should be made of water- and stain-resistant materials such as granite or other low-absorption back-primed stones, SW-grade brick, precast concrete, or cement plaster. Keep in mind that the period between the completion of the waterproofing and the installation of the facade may be longer than the membrane's ability to tolerate UV radiation. Permanent or temporary protection should be noted on the drawings.

D. Penetrations

Prepare a single drawing that shows all penetrations through the basement walls and slabs. This is usually called a point of entry (POE) drawing. It includes all mechanical and electrical penetrations, including electronic and utility piping. A POE drawing should contain a plan, show foundation wall elevations, and include all MEP piping with sleeve sizes indicated. These penetrations, more often than not, occur at varying elevations through the foundation wall. They may occur at the transition between pre-applied and post-applied waterproofing.

Many penetrations through the membranes applied to the pressure slab and the foundation wall are not always indicated on the drawings. Some, such as well points, may come as a surprise to you. All require special details.

1. Penetrations through foundation walls

Some penetrations through foundations are for utilities that require watertight sleeves when they are below the groundwater line. These include:

- sewer and water pipes
- conduits for power and electronic cables
- steam pipes
- geothermal pipes

Other penetrations through the wall include:

- anchors to tie the foundation back to an existing wall or rock face
- soil nails
- waler brackets (Figure 17.5)
- rakers, and
- diagonal corner braces

Figure 17.5 The secant piles will be cut back to the face of the soldiers and the waler brackets burned off. The membrane must be flashed around the bracket with metal to avoid damage from the torch

Sleeves are generally preferable to cast-in-place pipes and conduit for two reasons:

1. They permit independent movement of the pipes, thus averting the risk of fracture or distortion caused by movement in the structure.
2. They simplify the work of different trades by reducing the coordination required among them.

Inadequate space between penetrating pipes and conduits prevents the dispersion of aggregate in the concrete mix, creating honeycombed passages vulnerable to leakage. Careful coordination among the architectural, structural, mechanical, and electrical designers is required to assure adequate spacing and consequently solid concrete between pipes and conduits. Pipes and pipe sleeves should never be spaced closer than six inches apart to allow the installation of flashing without overlap.

Frequently, conduits are ganged, leaving very little space between them for proper flashing. These require precast concrete or heavy gauge sheet metal manifolds that are sealed to the foundation and flashed to the membrane (see Figure 17.6).

Figure 17.6 Sleeves for conduits are shown welded to a plate that is flashed into the membrane

Penetrations through single-ply, built-up, and liquid-applied membranes are flashed with strips of suitable bituminous or rubber membrane and special adhesive. Two or three successive applications of flashing are wrapped around the pipe, extending two to three inches beyond each other. A lead collar helps to absorb differential movement between the pipe and structure. A stainless-steel drawband secures the collar. In bentonite systems, penetrations are packed with troweled-on bentonite.

Conduits should also be spaced at least six inches apart in order to be properly flashed. Unfortunately, the space allotted in buildings is rarely sufficient to provide for such generous conduit spacing, and conduits are frequently ganged closer together. Where feeder and cable sleeves are nestled in a blockout in the wall they should be of sufficient diameter to provide not less than one inch between the sleeve wall and the conduit or cable. Concrete is packed in the blockout around the sleeves to secure them in place. The membrane must lap the joint between the wall and the block-out by at least two inches. A sheet metal form is fitted over the membrane and around the penetrations an inch from the face of the wall. It is then filled with a seal of pourable-grade, two-component urethane or the mastic component of bentonite systems (see Figure 17.6).

Flashing at penetrations in sleeves must accommodate differential movement between the slab and the pipe conduit. The space between the outer portion of the pipe and the sleeve is filled with oakum or a pre-compressed open-cell sealant. The inner space is sealed with a Link-Seal®[1] forced tightly against the pipe with bolts (see Figure 17.7). Large diameter pipes may require expansion joints. Link-Seals are also effective where other flashings may fail due to heat, movement, or chemical attack. The Link-Seals are relatively easy to replace.

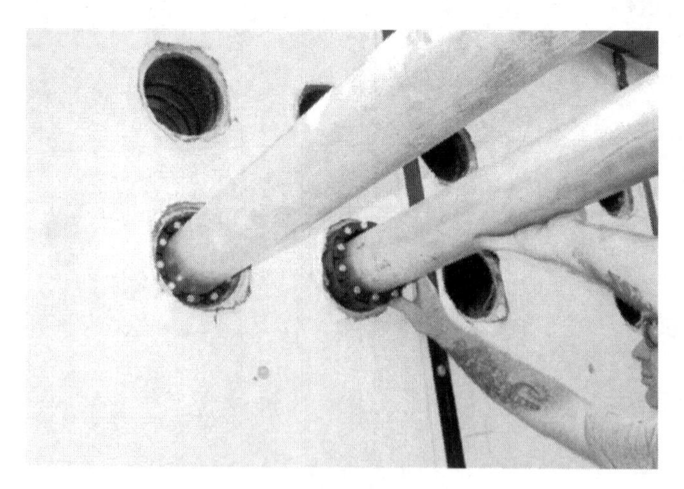

Figure 17.7 A LinkSeal® collar is fitted over the pipe in the annulus between the sleeve and the pipes and the nuts are tightened to form a watertight seal

Walers are often enclosed with sheet metal or plywood and waterproofed with the sheet membrane and the liquid flashing component.

2. Penetrations through the pressure slab

Leaking often occurs at penetrations. Leaks can be minimized by incorporating pits and piping into thickened or depressed pressure slabs. Otherwise, the more common method is to install pits and piping in sleeves. Conduits are often ganged in boxed sleeves. On plazas, single conduits – e.g., those feeding light fixtures – are cast into the structural slab.

Penetrations that frequently occur through the pressure slab include:

- caissons and pile caps that don't penetrate the slab, but interrupt the membrane under it. (Special foundations for construction cranes are not usually indicated in contract documents.)
- rock anchors
- lightning ground rods, as well as the building's electrical grounding systems
- well points
- drains and cleanouts
- pits; such as sumps, elevator pits.
- foot blocks or heel blocks
- hydraulic elevator pistons

Penetrations through the pressure slab can often be eliminated by depressing the slab and absorbing the sumps and pits.

While the basement is under construction and the dewatering system is operational, the weight of the foundation and pressure slab hold it in position. However, when the pumps are turned off, the basement will have a tendency to float and rise vertically. As the superstructure is constructed above, the imposed dead load will cause the basement to resettle. Where the basement is not constructed on rock, anticipate some movement, generally in the range of one to two inches. Movement is usually resisted by caissons, pin piles, or rock anchors to secure the pressure slab to the strata below. Those materials support the basement or restrain it from floating when the pumps are shut-off. Where they penetrate the pressure slab and mud mat, provision must be made to permit differential movement. Lightning ground rods, geothermal piping, and similar mechanical penetrations require special detailing.

The casing for a hydraulically powered elevator presents a special problem with pipe penetrations (see Figure 17.8). The top is cast into the concrete, and where the pipe is below groundwater level, it is temporarily filled with water to hold it in place.

The waterproofing membrane is flashed up the pipe to the top and concrete cast around it. A mortar plug with a chemical-conversion additive

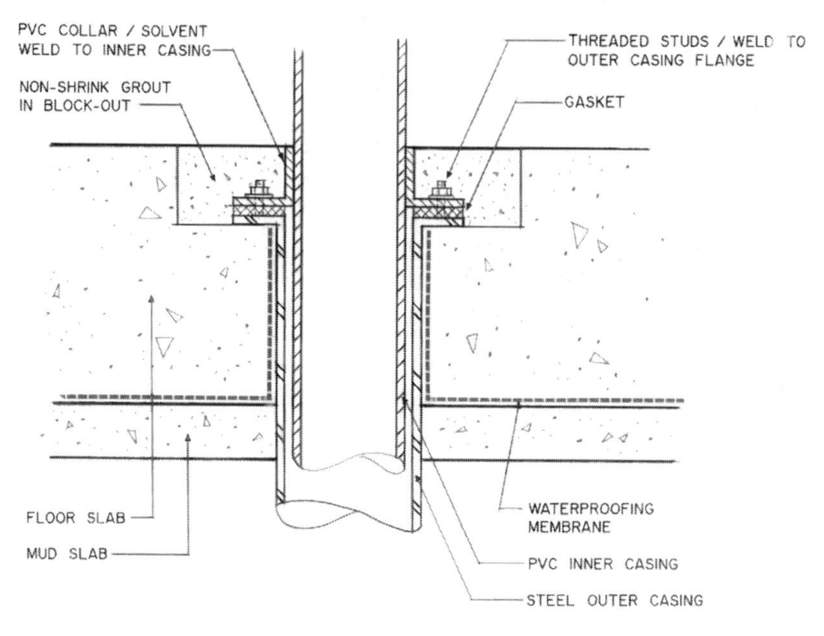

Figure 17.8 This detail is one solution to the problem of waterproofing the casing for a hydraulically powered elevator penetrating the slab on grade of the elevator pit

can be provided as a secondary seal at the top. If the elevator pit is waterproofed with negative-side cementitious or chemical-conversion coatings, the plug at the top becomes the primary seal and must be made deeper and wider.

E. Planters, bench, and equipment supports

Waterproofing membranes on horizontal surfaces should be carried under planters and supports without interruption. Such items should be installed on the concrete protection slab. Waterproofing of planters should be independent of the slab waterproofing (see Figures 13.6 and 13.7). When penetrations extend above the wearing surface, they must be carried at least eight inches above it and counterflashed with metal sleeves.

F. Details

The following details suggest methods that penetrations of pipes, conduits, rising walls, curbs and similar building components, as well as penetrations of the support of excavation system, can be flashed into the membrane while maintaining its water-tightness. They illustrate general design principles and should be modified for specific waterproofing systems. Keep in mind that waterproofing details are executed prior to installation of the field membrane. Understanding sequencing is critical to properly assembling the flashing components.

1. *Details at penetrations of the pressure slab with thermoplastic and modified bitumen sheets*

Figure 17.2c shows the waterproofing at the top of a caisson prior to casting the grade beam.

Rock anchors or mini-piles penetrating the pressure slab require flashing arranged to accommodate movement between the mud mat and the penetration.

Lightning grounding conductors are usually stranded cables. They cannot be made watertight when they penetrate waterproofing membranes because water flows around the strands. A coupling to transition from a stranded cable to a solid rod is required.

2. *Details at penetrations of the foundation wall with thermoplastic and modified bitumen sheets*

Single pipe penetrations are not a problem but ganged pipes are. Pipes are often ganged at transformer vaults where utility companies dictate the conduit spacing. Pipes or conduits are often so closely spaced that they are impossible to flash as illustrated in the waterproofing manufacturer's catalogue. The easiest solution is to provide an assembly comprised of plates with welded sleeves that can be placed in the form with its edges flashed into the wall. Care must be taken to avoid interrupting the rebar cages and to provide a frame to support the wall construction.

Pipes and conduits are inserted into the sleeves and made watertight with a series of rubber modular seals that are tightened with a wrench to form a watertight seal in the annulus around the pipe (shown in Figure 17.7).

Figure 17.9 shows a typical tie-back penetration. This may vary depending on whether the tie is a tension rod or cable strands. In either case, the tie will remain in place during the life of the building and must be flashed into the wall waterproofing system. This is usually accomplished by covering the tie with a prefabricated wood or metal box and flashing the box at the perimeter. One manufacturer markets a factory fabricated box. This one-size-fits-all unit may not be as flexible as a box fabricated by the waterproofing contractor.

A typical condition where the raker penetrates the pressure slab and under-slab waterproofing presents a difficult challenge to make the penetration water-tight. If possible, the SOE contractor should be encouraged to eliminate penetrating the membrane by:

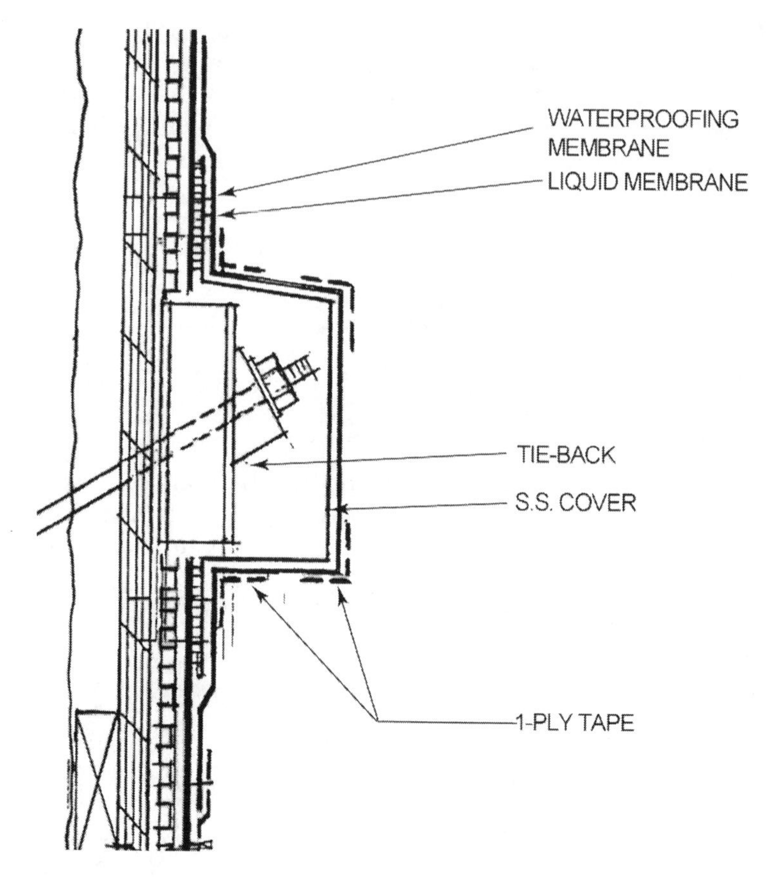

Figure 17.9 This detail shows one method of sealing tie-backs. Some membrane manufacturers provide their own cover

- raising the foot block so that the top is above the under-slab membrane;
- depressing the mud mat and thickening the pressure slab to absorb the foot block; or
- casting a section of the pressure slab and providing bolted angles to anchor the bottom end of the raker.

When these steps are not implemented, the penetration of the bottom of the raker must be flashed into the underslab membrane.

1. In Figure 17.10, the raker is boxed out with metal, and the flanges are flashed into the membrane. The void formed by the metal box can then be filled with quick-setting grout.
2. Alternately, the raker is flashed to the membrane using tapes and the liquid component of the membrane system.

Figure 17.10 The bottom of rakers that penetrate the pressure slab terminate in a foot or heel block that are often below the slab and penetrate the underslab waterproofing. One method is to flash it with the membrane, but a better method is to fit a sheet metal shoe around the raker and fill it with quick setting cement or a pourable sealer

Walers pose another problem. If they are held off from the earth retaining system, the stand-offs must be boxed-in. The challenge when cross-lot blocking or walers are to be removed is to avoid damage to the flashing from the heat of the cutting torch. This requires careful sequencing (and often repairs) after the steel members are burned off. Sometimes the enclosing boxes can be enlarged to conceal the damaged membrane.

Problems can often be resolved by staging the backfill or installing upper and lower tiers of rakers and walers. The earth retaining wall is prepared to hang the waterproofing by selectively removing upper and lower tiers and providing interim bracing with berms.

Another method is to locate the walers above an intermediate basement slab. The waterproofing is then carried up the foundation wall to the bottom of the waler and the slab cast against it. The waler is then removed and the waterproofing is continued. This method is frequently used on projects with tiers of cross-lot blocking.

G. Detailing checklist

1. Put waterproofing details, plus the extent of waterproofing, mud mats, drainage media, footing drains, and all other aspects of waterproofing

on architectural drawings or separate waterproofing drawings.

2. Check structural drawings for inclusion of cementitious cants and other concrete details required for the waterproofing.
3. Draw sections at a minimum scale of 3/4 inch = one foot, details at a minimum 1 1/2 inch = one foot, and penetrations, drains, plane changes, etc., preferably at 3 inches = one foot.
4. Use distinguishing indicators – e.g., alternating long and short dashes, alternating dots and dashes – for dampproofing versus waterproofing to avoid confusion when both appear in the same detail.
5. Make sure that plumbing (or site) drawings show footing drains in plan and connections to site utilities.
6. Include details for support of excavation components that penetrate the membranes.

Note

1 Link-Seal is manufactured by GPT, Denver, CO.

Appendix A
SI conversion factors

Table A.1 SI conversion factors

To convert from	*To*	*Multiply by*
°F	°C	0.556 (°F–32)
°C	°F	1.8 (°C+32)
mm	in.	0.03937
in.	mm	25.400
g/m^2	lb/square (100 ft^2)	0.0205
lb/square (100 ft^2)	g/m^2	48.83
pcf (lb/ft^3)	kg/m^3	16.02
kg/m^3	pcf	0.0624
psf ($lb/ft.^2$)	kg/m^2	4.882
kg/m^2	psf	0.205
psi (lbf/in^2)	kPa	6.895
kPa	psi	0.1450
lbf/in.	kN/m	0.1751
kN/m	lbf/in.	5.710
Perm (vapor permeance) [grain/($ft^2 \cdot h \cdot in.Hg$)]	ng/($m^2 \cdot s \cdot Pa$)	57.45
R [°F/(Btuh · ft^2)]	R [K/W · m^2)]	0.176
R [K/(W · m^2)]	R [°F/Btuh · ft^2)]	5.678
U [Btuh/(ft^2 · °F)]	U [W/m^2 · K)]	5.678
U [W/m^2 · K)]	U [Btuh/ft^2 · °F)]	0.176
grain	g	0.0648
Btu	J	1055
gal/square (100 ft^2)	L/m^2	0.408

Key to abbreviations:

°F = degrees Fahrenheit
°C = degrees Celsius
K = Kelvin (= °C + 273.15)
Btuh = Btu per h
lbf = pound · force
N = newton (= 0.2248 lbf)
kN = kilonewton = 10^3N
m = meter
mm = millimeter
g = gram
kg = kilogram (= 10^3g)
ng = nanogram (= 10^{-6} kg = 10^{-9} g)
Pa = pascal (= N/m^2)
kPa = 10^3 Pa
W = watt (= 3.414 Btuh)

Appendix B
Plaza surfacing

Plazas are surfaced with cast-in-place concrete, natural stone, tile, precast masonry units, or wood planking. The material is installed in a setting bed or on pedestals with open joints. Unlike pavers used for ballast in protected membrane roof (PMR) or inverted roof membrane assembly (IRMA) systems, plaza materials are intended for pedestrian and light vehicular traffic and are not installed directly on a membrane (but may be installed on the insulation above it).[1] Whereas pavers used for ballast are selected for resistance to wind-uplift, pavers for plazas are selected for resistance to static and dynamic loads, petroleum spills, and impact.

Open-joint pavers are made from one of four materials:

1. Precast concrete.
2. Natural stone.
3. Prefabricated wood decking.
4. Porcelain.

Precast concrete is hydro-pressed or wet-cast in molds, in units weighing 15 to 25 pounds per square foot, in square or rectangular shapes ranging from 12 by 14 inches to 30-square inches. Thickness generally ranges from 1 1/4 to 2 1/2 inches, but sometimes up to four inches. Faces are available in varied textures and a wide range of colors, exposed aggregates, and simulated tile. Backs are usually flat, sometimes with lugs cast in at corners and centers. Maximum allowable water absorption is five percent. Compressive strength is around 7,000 psi. Weight loss after 100 freeze/thaw cycles per ASTM C1262 test should not exceed one percent. There are no ASTM standards for these units.

Natural stones with water-absorption rates below one percent include granite (per ASTM C615); marble, travertine (per ASTM C503) flagging of slate (per ASTM C629); and bluestone and other fine-grained quartz-based stone (per ASTM C616). High-density limestone with maximum water absorption of three percent (per ASTM C568) is also included. For stones with less flexural strength, thickness ranges from 1 1/2 to four inches. Sizes and shapes vary, but rectangular and square stones predominate for pedestal installations. Faces are available in a wide range of textures including thermal (for granite), natural cleft for flagging, sand blasted, and smooth. Backs are usually gauged (smooth or honed).

Prefabricated wood decking units are shop-fabricated from a variety of wood species. Certain species – e.g., teak, ipe, mahogany, redwood, and cedar – are inherently weather resistant. Others usually require pressure-preservative treatments. Panels are assembled from boards set flat or on edge with open or tight joints. They are screwed to frames or bolted together.

Units should be heavy enough to resist blow-off. Dimensional stability and code requirements for fire resistance are other considerations.

Porcelain pavers are manufactured with a non-slip porcelain top layer bonded to a ceramic base. They are produced in 24-inch by 24-inch and 24-inch by 48-inch shapes with thicknesses of $1\ ^{1}/_{16}$-inch to $1\ ^{5}/_{16}$-inch respectively. Porcelain pavers are fire-proof, stain and chemical resistant and can support loads in excess of 2000 pounds.

Paver materials for closed-joint construction include:

- precast concrete
- natural stones (listed above)
- clay bricks
- concrete-masonry units
- tile (ceramic and quarry)
- rubber

Ceramic tile in locations subject to freezing should have maximum water absorption of three percent per ASTM C373, plus classification as *resistant* per ASTM C1026 test. All units should have a minimum coefficient of friction (wet and dry) of 0.6 per ASTM C1028. Bricks should be brick pavers per ASTM C902, never facing or common bricks.

Precast units, clay bricks, and stones are laid dry on pedestals or in sand or mortar beds. Bricks may also be installed in bituminous setting beds. Pedestals create a level surface, whereas sand or mortar beds often follow the contour of the suspended slab below. A typical setting bed is 1:3 to 1:6 Portland cement and sand mixed to a damp consistency, and tamped. The back of the paving unit or the top of the setting bed is gauged with cement and water. The unit is then laid dry with a tight joint and tamped. The joints are caulked or the same mix is swept dry into the joints with the excess cleaned with a light spray of water. Polymeric sand is frequently specified for filling sand swept joints. It resists displacement and organic growth.

Rubber pavers are manufactured from EPDM and are for use on plazas for athletic facilities. They are 24-inches by 24-inches with interlocking edges and in thicknesses ranging from $1^{3}/_{4}$-inches to $3^{3}/_{4}$-inches depending on the height of fall resistance required.

Prefabricated rubber and plastic pedestals are manufactured in a variety of designs. They provide a level surface over undulating substrates (see

Figure B.1). Heights are adjusted by stacking, interlocking, and shimming; or by rotating male and female threaded units. Some units require stabilizing with mortar-filled cores. However, capillary draw can bring moisture to the top of the pedestals and stain the pavers (see Figure B.2). Short pedestals can be installed on stacked squares of insulation or small pavers. However, this technique creates problems when dismantling is required for maintenance or repair. All units contain tab spacers to create uniform joints and shims to adjust pavers individually.

Pedestals are often installed atop the insulation layer, but may be installed on the membrane, provided protection board, or several layers of single-ply sheets interposed between the pedestal and the membrane. Installing pedestals on the membrane avoids the need to accommodate compressive creep inherent in all plastic foams. But it complicates the insulation installation and creates thermal bridges.

Paving brick, quarry tile, and ceramic tile set in mortar beds are also used for wearing surfaces.

Sand-setting beds can be installed over a concrete protection slab or filter cloth on rigid insulation. Brick pavers in sand-setting beds are more successful than mortar setting beds, according to Clayford T. Grimm, the late masonry engineering expert. Eccentric loading over small areas can cause

Figure B.1 Pavers being installed on the plaza of a hotel over guest rooms. The membrane is loose-laid PVC on a lightweight concrete fill sloped to drain. The pedestals adjust to the slope to provide a level surface for the promenade tables and chairs

Figure B.2 These granite pavers were installed over hollow pedestals filled with mortar. The poorly compacted mortar filling enabled moisture on the plaza to rise capillarity and discolor the granite paver joints

high stresses in thicker sand beds that promote settling, rocking, and loosening; according to Estensorro and Perenchino.[2] They recommend sand beds of half-inch to one-inch thickness, with an absolute maximum of three inches. A one-inch thickness is often inadequate to accommodate subgrade variations and to allow for raising the level of the substrate at expansion joints. To solve this problem, a one-inch sand bed can be installed over filter cloth on three- to six-inch strata of coarse gravel or a drainage composite. This provides the additional benefit of preventing frost heaving in areas subject to subfreezing temperatures. Sand should be coarse and single size, gap-graded (not like mortar sand) for maximum porosity.

Slate and other flagstones not exceeding 18 by 18 inches can also be set in this manner. Larger units should be installed on tamped damp mortar beds with low cement/sand ratios, such as 1:3 and 1:6.

In climates subject to freezing, mortar beds for small stones and tile should contain acrylic additives to prevent freeze/thaw deterioration.

Mortar setting beds over insulation should be installed over a filter sheet and reinforced with hexagonal mesh.

Sand-setting beds provide sub-surface drainage, but mortar setting beds must be installed over porous fill or drainage composites.

For construction details and specifications for ceramic and quarry tile refer to the Tile Council of America Ceramic Tile's *The Installation Handbook, 1997, Roof Deck Membrane F103-94.*

Notes

1 Concrete pavers for ballasting PMR systems that are installed dry on a roof membrane. They are produced by concrete-masonry unit manufacturers on block-making equipment. Weighing approximately 12 psf, these pavers are square or rectangular, varying from a minimum 6 × 12 to a maximum 12 × 18 inches. Thickness ranges from 1 1/2 to 2 5/8 inch. Most have square edges, but units are also made with beveled edges and interlocks for greater wind-uplift resistance. Faces are available in a variety of textures and a limited range of colors. Backs are flat, corrugated and grooved. Larger units have lugs cast in the corners and center to permit insulation drying. These units have absorption rates in the 10% range and compressive strengths of 2,600 to 3,000 psi. Weight loss after 100 cycles of freeze/thaw testing per ASTM C1262 should not exceed one percent.

These units are not recommended for a wearing surface subject to normal pedestrian or light vehicular traffic. The lone exception is interlocking concrete-masonry pavers that conform to ASTM C936, which may be used when set in a sand bed.

2 Estenssoro, Luis F., and William F. Perenchio, Failures of Exterior Plazas, *Construction Specifier*, January, 19911/91 pp 87–92.

Appendix C

Waterstops

Waterstops are required in both horizontal and vertical "cold" joints – i.e., construction and expansion joints – subject to hydrostatic pressure. They are embedded in the concrete, spanning across the joint to form a barrier to passage of water. With negative-side waterproofing, waterstops are a primary part of the system; with positive-side and bentonite systems, they provide a secondary line of protection.

The original waterstops were simple metal bars spanning across the joint, with one side greased to break the bond with the concrete and prevent stress concentrations in the waterstop metal at the joint. Despite the greasing precaution, these simple waterstops proved inadequate for moving joints. They were replaced with thin copper sheets, bent to form flanged Vs. These, too, proved inadequate. Cyclical movement work-hardened the metal and promoted fatigue failure. Rubber dumbbells were the next unsuccessful attempt. They leaked when water migrated around their ends or when the ends pulled free of the concrete.

Contemporary waterstops are formed from PVC, HDPE, TPV or rubber derivatives, hydrophilic bentonite, and aliphatic/butyl as well as non-swelling mastic strips. They are available in a variety of shapes, including serrated U and V shapes, labyrinth, center bulb, and ribbed dumbbell (see Figure C.1). Center bulb types are used for expansion joints in walls. Where these are intended to be the sole water barrier, they must be designed for resistance to hydrostatic pressure and water migration around their ends. For design parameters, consult the Corps of Engineers, "Waterstops and other Joint Materials," EM1110-2-2102 and CRDC-C-572-74.

U-shaped center bulbs with ribbed flanges have increased the flexibility and reduced infiltration of water around the flanges. However, they have not overcome the greatest weakness of bulb-type waterstops: inability to achieve a perfect weld at end joints and corners (see Figure C.2).

Units are tied to reinforcing and spliced by site-welding. Great care is required to insure proper positioning and to preserve the integrity of splices. Care must continue during concrete casting operations, to insure encasement in solid, void-free concrete and to prevent displacement by the concrete mix.

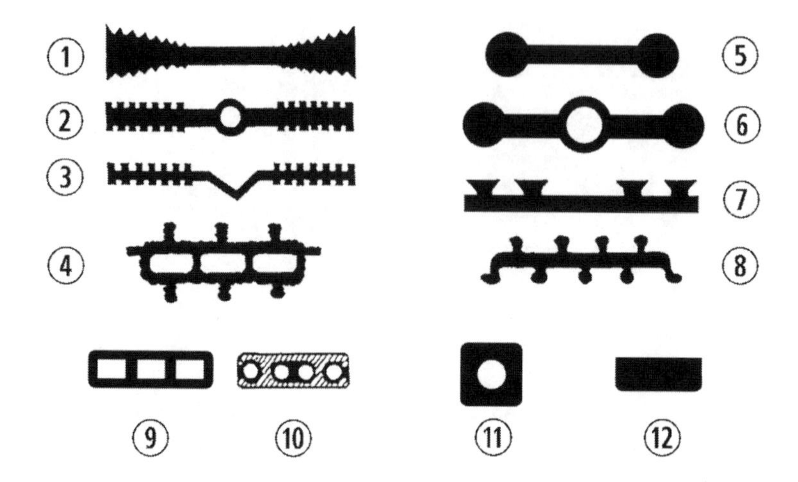

Figure C.1 Basic types of waterstops: (1) serrated; (2) serrated dumbbell; (3) serrated "V"; (4) cellular; (5) dumbbell; (6) center bulb dumbbell; (7) on-grade; (8) labyrinth; (9) modified; (10) chloroprene; (11 and 12) butyl/bentonite

Figure C.2 Serrated plastic waterstop in a vertical cold joint

Partly because of the difficulty in assuring this required degree of care, other types of waterstops are gradually replacing the dumbbell and tube-type PVC waterstops. These new types are composed of butyl rubber, bentonite-impregnated butyl, and proprietary plastics (for static joints).

Formed as square or rectangular rods, modern waterstops are relatively easy to bend and splice. Installation is equally simple: units are nailed or adhered to the concrete or to the forms, a practice that eliminates form-splitting. The bentonite/butyl compositions have the additional advantage of permitting installation on damp substrates. However, the longevity of hydrophilic-type joints and of bentonite when a waterstop is under a continuous high hydrostatic head is not yet established.

Mastic waterstops are an economical alternative to hydrophilic expanding waterstops. One advantage is that they are not susceptible to pre-hydration. They are designed to be adhered in a primed keyway on one side of the joint of a previously cast concrete foundation or slab. When the adjacent concrete is cast around it, the heat of hydration of the cement causes the waterstop to adhere to the concrete. However, their performance is inferior to hydrophilic waterstops.

The hydrophilic and mastic types of waterstops are easier to install and less likely to be displaced by subsequent concrete pours than the extruded plastic types, but they are still subject to failures from improper installation and damage. For more information refer to Stacy Byrd's exhaustive article, "Understanding Waterstops," in the July 2015 issue of *The Construction Specifier*.

A belt-and-suspenders approach is often warranted when an interior space is particularly sensitive to water infiltration or joints are likely to fail. This is accomplished by installing special hoses in the joints in tandem with preformed waterstops during the casting operation and injecting urethane or acrylate esters after the concrete has cured, which typically takes 30 days. The hoses are fabricated from a hard rubber compound and are perforated on four sides. The perforations are covered with strips of rubber and the hoses are then wrapped in fabric.

When the injection chemicals are pumped through the hoses, the rubber strips are lifted sufficiently to permit the chemicals to disperse through the joint and swell. The hoses are then back-washed with water and can be reused several times, if required. Some manufacturers provide paired hoses.

The hoses are installed in 40- to 60-foot lengths and the ends are lapped. Hoses are secured to the previously cast concrete element by special clips. They are terminated at injection boxes with covers, similar in size and shape to conventional electric junction boxes, and installed in the walls (see Figure C.3).

Injection can take place any time after the concrete has cured for 30 days. It may be deferred for weeks or even years, if the installed conventional waterstops are successfully preventing water infiltration. It can also be used repeatedly when water pressures fluctuate or seismic forces or settlement

Figure C.3 An injectable hose installed in a construction joint keyway. The hose may be left open and filled with chemical grout at a later date or injected as soon as the concrete has aged 28 days

causes joints to open and leaks to occur. Injection can be localized to individual circuits of hoses to suit the leaking area.[1]

Note

1 Installation Failures Reprinted from *Understanding Waterstops* in the July 2015 issue of *The Construction Specifier* by Stacy Byrd
The problem with waterstops is their susceptibility to improper installation or damage during the concrete placement. The following list illustrates some of the many potential installation failures for waterstops.

- dumbbell or ribbed center-bulb roll ends overlapped but not welded or spliced together;
- installed too close to steel reinforcement;
- dumbbell splices glued together with sealant; not welded;
- strip waterstop installed with concave gap (void) under it;
- polyvinyl chloride (PVC) transition glued together-no fabrication part;
- PVC welded on the edge only and not fully across its profile thickness;
- poorly consolidated concrete adjacent to waterstop;
- overheated burnt or charred thermoplastic welds;
- dumbbell or ribbed center-bulb product not centered in joint;

- dumbbell not properly tied into reinforcement so it shifted during concrete pour;
- hole cut in flange of dumbbell to pass the rebar through;
- overlapping, not butting, hydrophilic strip roll ends;
- flange of dumbbell cut narrower to fit around reinforcing steel;
- misaligned ribs or centerbulb at splice;
- concrete extending on flange not removed from waterstop prior to second pour; and
- hydrophilic-strip waterstops installed only with fasteners

Appendix D

Abstracts of ASTM standards relating to waterproofing

The following ASTM Standards apply to materials and practices for using them commonly found in waterproofing specifications. The abstracts have not been authorized by ASTM, and may not be complete or current as you read this. You should review any standard before referencing it in a project manual.

C578 Standard Specification for Rigid, Cellular Polystyrene Thermal Insulation. Covers the types, physical properties, and dimensions of cellular polystyrene insulation boards. Includes boards formed by expansion of polystyrene resin beads or granules in a closed mold (also called EPS), and boards formed by expansion of polystyrene base resin in an extrusion process (also called extruded polystyrene). Eleven types are listed, with densities ranging from.70 to 3.00 psf and compressive strengths ranging from 5 to 100 psi. The most commonly used types for waterproofing listed with their density (psf) and compressive strength (psi) are: VI (1.60 psf, 40 psi) and VII (2.20 PSF [60 psi]). First published in 1965.

C755 Standard Practice for Selection of Vapor Retarders for Thermal Insulation. Outlines factors to be considered, describes design principles and procedures for vapor retarder selection, and defines water vapor transmission values appropriate for established criteria. Although intended for vapor retarders for thermal insulation, the standard's discussion of vapor retarder materials is useful. First published in 1973.

C836 Standard Specification for High-Solids-Content, Cold Liquid-Applied Elastomeric Waterproofing Membrane for Use with Separate Wearing Course. Describes required properties and test methods for a cold, liquid-applied elastomeric-type membrane (one- or two-component) for waterproofing building decks subject to hydrostatic pressure in building areas to be occupied by personnel, vehicles, or equipment. This specification applies only to a membrane system that will have a separate wearing or traffic course applied over it. Physical requirements include hardness, weight loss, low-temperature flexibility and crack-bridging, film thickness on vertical surface, adhesion-in-peel after water immersion, and extensibility after heat aging. First published in 1976.

C898 Standard Guide for Use of High Solids Content, Cold Liquid-Applied Elastomeric Waterproofing Membrane with Separate Wearing Course. Describes the use of a high-solids-content, cold liquid-applied elastomeric waterproofing membrane in a waterproofing system subject to hydrostatic pressure for building decks over occupied space where the membrane is covered with a separate wearing course. Contains design considerations for substrates, membranes, protection courses, drainage systems, and wearing courses. First published in 1978.

C981 Standard Guide for Design of Built-Up Bituminous Membrane Waterproofing Systems for Building Decks. Describes the design and installation of bituminous membrane waterproofing systems for plaza deck and promenade construction over occupied spaces of buildings where covered by a separate wearing course. Contains design considerations for substrates, membrane components, protection board, drainage, insulation, protection, and wearing courses. An excellent primer for plaza design. First published in 1983.

C1127 Standard Guide for Use of High Solids Content, Liquid-Applied Elastomeric Waterproofing Membrane with an Integral Wearing Surface. Describes the design and installation of cold-applied elastomeric waterproofing membrane systems that have an integral wearing surface. The system consists of a reinforced concrete substrate coated with a primer, base and top coats, and embedded aggregate. The standard contains installation details. First published in 1989.

D41 Standard Specification for Asphalt Primer Used in in Dampproofing and Waterproofing. Defines the minimum requirements for asphaltic primer use with asphalt in waterproofing below or above ground level and applied to concrete and masonry surfaces. First published in 1917.

D173 Standard Specification for Bitumen-Saturated Cotton Fabrics Used in Roofing and Waterproofing. Defines the minimum requirements for bituminized cotton fabric composed of woven cotton cloth waterproofed with either asphalt or coal-tar pitch. First published in 1923.

D226 Standard Specification for Asphalt-Saturated Roofing Felt Used in Roofing and Waterproofing. Defines the minimum requirements for two types of asphalt-saturated, but not coated, felts: No. 15 and No. 30. Felts may be perforated or non-perforated. First published in 1925.

D227 Standard Specification for Coal-Tar-Saturated Organic Felt Used in Roofing and Waterproofing. Defines the minimum requirements for coal-tar-saturated organic felts for use with coal-tar pitch. First published in 1925.

D449 Standard Specification for Asphalt Used in Dampproofing and Waterproofing. Defines three types of asphalt suitable for use as a plying or mopping cement in construction of a membrane waterproofing system. Type I is a soft adhesive "self-healing" asphalt that flows easily under a mop and is suitable for use below ground level under uniformly moderate temperature conditions. Type II is somewhat less susceptible to temperature, with good adhesive and "self-healing" properties for use above ground level where not exposed to temperatures exceeding 125° F. Type III is least susceptible to temperature, with good adhesive properties for use above ground level where exposed on vertical surfaces in direct sunlight or at temperatures above 125° F. First published in 1937.

D450 Standard Specification for Coal-Tar Pitch for Roofing, Dampproofing, and Waterproofing. Defines three types of coal-tar pitch suitable for use in construction of built-up roofing, dampproofing and membrane waterproofing systems. Type I is suitable for built-up roofing, dampproofing, and membrane waterproofing systems. Type II is suitable for dampproofing and membrane waterproofing systems. Type III is suitable for built-up roofing, dampproofing and membrane waterproofing systems, but with fewer volatile components than Type I or II have. First published in 1937.

D698 Standard Test Methods for Laboratory Compaction Characteristics of Soil Using Standard Effort (12 400 ft-lbf/ft3 (600 kN-m/m3). See also **D1557 Standard Test Methods for Laboratory Compaction Characteristics of Soil Using Modified Effort (56,000 ft-lbf/ft3 (2,700 kN-m/m3).**

D1056 Standard Specification for Flexible Cellular Materials – Sponge or Expanded Rubber. Covers two types and four classes of rubber. Type I is open and Type II is closed cell. : Class A is not oil resistant, Class B is low swell from oil and Class C is medium swell. Class D is resistant to temperature extremes. Each class is available in three grades with various compression deflection ranges. They range in increments from 0–.5, .5–2 and 5–17 psi. First published in 1949.

D1327 Standard Specification for Bitumen-Saturated Woven Burlap Fabrics Used in Roofing and Waterproofing. Defines the composition of woven burlap cloth saturated with either asphalt or refined coal-tar. First published in 1954.

D1668 Standard Specification for Glass Fabrics (Woven and Treated) for Roofing and Waterproofing. Covers three types of glass fabrics composed of woven glass cloth treated with either asphalt or coal-tar pitch or an organic resin. Type I is asphalt-treated, Type II is coal-tar pitch-treated, and Type III is organic resin treated. First published in 1959.

D1752 Standard Specification for Preformed Sponge Rubber Cork and Recycled PVC Expansion Joint Fillers for Concrete Paving and Structural Construction. Covers three types: Type I (sponge rubber), Type II (cork) and Type III (self-expanding cork). First published 1960.

D2178 Standard specification for Asphalt Glass Felt Used in Roofing and Waterproofing. Covers three types of glass felts impregnated with asphalt, Types III (8.4 lb/110 sq. ft.), IV, and VI (both 6.0 lb/100 sq. ft.). Types III and IV contain pinholes. Type VI is perforated with 1/8-inch in diameter holes. First published in 1963.

D4397 Specification for Polyethylene Sheeting for Construction, Industrial, and Agricultural Applications. Covers polyethylene sheets up to 10 mils thick. Physical characteristics include color and finish, impact resistance, mechanical properties, reflectance, luminous transmittance, water vapor transmission from .014 to 1.4 grains/24 hr/100 sq. ft. and permeance from .076 to .76 perms.

D4434 Standard Specification for Poly (Vinyl Chloride) Sheet Roofing. Covers reinforced flexible sheets made form poly (vinyl chloride) resin for use in single-ply roofing membranes exposed to weather. Type II, Grade 1 is internally reinforced. Type II, Grade 2 is externally reinforced with fabric. Type III is internally reinforced with fabric and may also have a fabric backing. Type IV is internally reinforced with fabric and may also have a fabric backing at least .036 inches thick. Types II and III are .45 inches thick. Type IV is .91 inches thick. First published in 1985.

D4637 Standard Specification for EPDM Sheet Used in Single-Ply Roof Membrane. Covers non-reinforced, fabric- or scrim-reinforced, and fabric-backed vulcanized rubber sheets made from EPDM for use in roofing. Type I is non-reinforced; Type II is internally reinforced scrim (or fabric); Type III is fabric-backed. First published in 1987.

D5295 Standard Guide for Preparation of Concrete Surfaces for Adhered (Bonded) Membrane Waterproofing Systems. Provides recommendations for preparation of new concrete surfaces before application of adhered (bonded) waterproofing. Discusses adhesion inhibitors, repair of surface defects, surface preparation, and evaluation by field tests. First published in 1992.

D5385 Standard Test Method for Hydrostatic Pressure Resistance of Waterproofing Membranes. Provides a laboratory test for measuring the hydrostatic resistance of a waterproofing membrane and can be used to compare the hydrostatic resistance between different membranes. First published in 1993.

D5474 Standard Guide for Selection of Data Elements for Groundwater Investigations. Covers the selection of data elements for documentation of groundwater sites. Excellent list of documents relating to subsurface investigations of soil, rock, and groundwater. First published in 1993.

D5665 Standard Specification for Thermoplastic Fabrics Used in Cold-Applied Roofing and Waterproofing. Covers thermoplastic fabrics such as polyester, polyester/polyamide bi-component, or composites with fiberglass or polyester scrims for use in cold-applied waterproofing. Type I is polyester spunbonded without resin, un-needled. Type II is polyester spunbonded without resin, needled. Type III is polyester mat plus fiberglass scrim with resin. Type IV is polyester core/polyamide sheath bicomposite spunbonded. Type V is polyester mat with polyester stitching. Type VI is polyester mat plus polyester scrim with resin. First published in 1995.

D5726 Standard Specification for Thermoplastic Fabrics Used in Hot-Applied Roofing and Waterproofing. Covers thermoplastic fabrics such as polyester or polyester/polyamide bi-component; or composites with fiberglass or polyester scrims for use in hot-applied waterproofing. Type I is polyester spunbonded with resin, un-needled. Type II is polyester spunbonded with resin, needled. Type III is polyester mat plus fiber glass scrim and resin. Type IV is polyester core/polyamide sheath bicomposite spunbonded. First published in 1995.

D5843 Standard Guide for Application of Fully Adhered Vulcanized Rubber Sheets Used in Waterproofing. Provides information for application and protection of fully adhered EPDM, butyl, and neoprene vulcanized rubber sheets on concrete substrates. First published 1996.

D5898 Standard Guide for Standard Details for Adhered-Sheet Waterproofing. Illustrates details for typical conditions encountered in adhered-sheet waterproofing membranes on below-grade structures and plazas. Contains drawings and explanatory notes for detailing terminations, penetrations, expansion joints and drains. First published in 1996.

D5957 Standard Guide for Flood Testing Horizontal Waterproofing Installations. Describes a method for testing the water-tightness of waterproofing installations applied to horizontal surfaces with a slope of up to two percent. The membrane is dammed and covered with one to four inches of water for 24 hours. First published in 1996.

D6134 Standard Specification for Vulcanized Rubber Sheets Used in Waterproofing Systems. Covers two types of rubber: Type I, EPDM; and Type II, Butyl (IIR). First published in 1998.

D6135 Standard Practice for Application of Self-Adhering Modified Bituminous Waterproofing. Outlines general procedures for application of self-adhering modified bitumen waterproofing systems in new installations. First published 1998. Withdrawn in 2014 because it was not recently updated.

D6769 Standard Guide for Application of Fully Adhered, Cold-Applied, Prefabricated Reinforced Modified Bituminous Membrane Waterproofing Systems. Contains application requirements for built-up modified bitumen membranes applied to below-grade basement walls and slabs. First published in 2002.

D7492 Standard Guide for Use of Drainage System Media with Waterproofing Systems. Contains guidelines for selection and installation of drainage system media used with waterproofing systems. First published in 2012.

D7877 Standard Guide for Electronic Methods for Detecting and Locating Leaks in Waterproofing Membranes. Describes procedures for using electrical conductance measurement methods to locate leaks in exposed or covered waterproof membranes. First published in 2014.

E154 Standard Test Methods for Water Vapor Retarders Used in Contact with Earth Under Concrete Slabs, on Walls, or as a Ground Cover. Describes various test methods for determining the properties of flexible membranes to be used as vapor retarders. Properties include water vapor transmission, tensile strength, low-temperature bending, resistance to puncture, plastic flow deterioration from soil organisms and substances, soil poisoners, UV exposure, and flame spread. First published in 1959.

E1643 Standard Practice for Selection, Design, Installation, and Inspection of Water Vapor Retarders Used in Contact with Earth or Granular Fill Under Concrete Slabs. Covers procedures for installing flexible, prefabricated sheet membranes under concrete slabs on ground. Describes placement and protection. Contains a field checklist and illustrations. Appendix covers pre-design considerations for landscaping and irrigation, subgrade design and specification, and base and protection courses. First published 1994.

E1745 Standard Specification for Plastic Water Vapor Retarders Used in Contact with Soil or Granular Fill Under Concrete Slabs. Covers non-bituminous, flexible, preformed sheet materials to be used as vapor retarders under concrete slabs on ground. Lists three classes with values for permeance, tensile strength, puncture resistance, and flame spread. First published in 1996.

F710 Standard Practice for Preparing Concrete Floors to Receive Resilient Flooring. (supersedes F1907 Standard Guide to Methods of Evaluating Moisture Conditions of Concrete Floors to Receive Resilient Floor Coverings). Covers the acceptance of concrete slabs to receive vapor retardant finishes by six tests: polyethylene sheet test, mat test, electrical resistance test, RMA test, primer or adhesive strip test, and humidifier test. First published in 1998.

E1907 Standard Guide to Methods of Evaluating Moisture Conditions of Concrete Floors to Receive Resilient Floor Coverings. Covers the acceptance of concrete slabs to receive vapor retardant finishes by six tests: polyethylene sheet test, mat test, electrical resistance test, RMA test, primer or adhesive strip test, and humidifier test. First published in 1998.

Appendix E

Expansion joints

Detailing expansion joints in below-grade construction is a great challenge. Expansion joints – more accurately called movement joints – are located between structurally independent building segments. Expansion joint movement is reversible and cyclical, and movements can continue throughout a building's service life. Expansion joints are thus distinguished from control or construction joints, where most movement is contraction (generally from the concrete's initial settlement and drying shrinkage), which consequently widens joints.[1,2]

Although more than 20 different stresses can cause movement, the most common for below-grade structures are from temperature changes, settlement, flotation, and seismic forces. Other stresses, such as those produced by creep, will usually be accommodated by joints designed for the aforementioned stresses. Where movement from moisture changes can be significant in above-ground structures, below-ground moisture content differentials in soil are relatively small. Moreover, the moisture content of concrete protected by membrane waterproofing is nearly constant.

Thermally induced movement is generally linear where framing materials and systems do not change. Thermally induced movement will be greater in structural slabs supporting plazas than in building components below-grade where temperature differentials are small. However, differentials can be significant in structures that are exposed for long periods of time before backfilling.

Differential movement can be expected from settlement, flotation, and seismic forces as well as changes in framing, interior environment, relative stiffness, and similar conditions. The design to accommodate movement from seismic forces below-ground is generally similar to that above-ground. Movement from settlement and flotation will be a function of the soil and water table characteristics. Their magnitudes should be determined by structural and geotechnical engineers. Settlement is generally non-reversible, but flotation can vary with rise and fall of the groundwater level. This can be significant where, for example, a lightweight tunnel joins a building foundation in tidal areas.

Properly designed expansion joints permit independent movement of the building elements in three dimensions on each side of the joint. Joint size and location should be determined by a structural engineer. Where slopes are integral with structural slabs, expansion joints should be located at the high points.

Joints designed for reversible movement will experience their maximum movement during and shortly after construction when backfilling is completed, but prior to actuating the HVAC systems, which will minimize thermal movement by maintaining a narrow range of temperature and humidity conditions. Initial significant non-reversible movement can be expected when earth-covered slabs are backfilled, heavy plaza wearing courses are installed, dewatering systems are deactivated, and foundations are loaded with the superstructure.

An expansion joint is a clear path between interior and exterior. Use them judiciously in your designs. Expansion joints are usually required where below-grade pedestrian and utility tunnels enter basements and where subsurface additions are constructed (see Figure E.1). Joints are also required in large plazas and where plazas adjoin rising walls. Both underground pipe and pedestrian tunnels are usually cast after the main foundation wall is constructed. When the tunnels are covered with earth, joint between the tunnel and foundation wall will usually experience deflection. This is more severe when the foundation is anchored to caissons or piles. After the tunnels are covered, flotation reverses the movement.

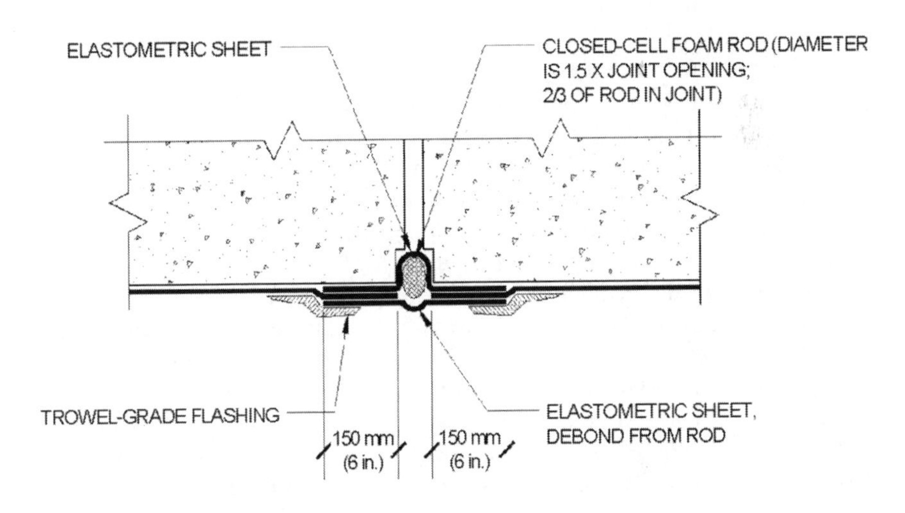

Figure E.1 This expansion joint in the wall of a tunnel close to where it joins the foundation can be used for the top and sides of the tunnel. The joint should be placed on a stub wall about 12 inches from the face of the foundation. (ASTM International)

In addition to perimeter expansion joints, large plazas may require structural expansion joints in the plaza field, generally spaced 100 to 200 feet apart. (These joints are usually designed by the structural engineer.) Wearing courses, which are exposed to temperature changes at least 10 times greater than those experienced by the insulated, underlying structure, require a correspondingly greater number of expansion joints extending only through the wearing course section. A wearing course expansion joint must coincide with each structural expansion joint. At these locations, the expansion joint width at the wearing course must approximate the width of the structural expansion joint. The single exception to this rule is where the wearing course consists of loose-laid units.

An expansion joint should be designed as a straight joint with its sides normal to the plane of the slab and foundation. It should be covered with a flexible sheet that can absorb movement without dependence on its elastic properties. Early joints were fabricated from folded copper, but they often failed from fatigue and work hardening. Today's joint covers are usually neoprene or silicone, looped over a rubber tube or proprietary support. A factory-fabricated unit with an elastomeric sheet bonded to metal flanges is also used.

You should avoid a notoriously defective type of expansion joint. This sealant-filled joint is shielded by an elastomeric strip. If a joint is truly dynamic, the width required for a sealant with ± 25 percent movement capability can exceed two inches. Moreover, if the sealant fails, it is inaccessible for repair. With liquid-applied and self-adhering waterproofing systems, the resistance to stress is dependent on the shear strength between the membrane and the joint cover.

An exception to this is an expansion joint filler composed of a silicone or polysulfide coated modified acrylic impregnated foam. It is designed to perform in submersible conditions with a five-foot head.

Though not subject to the wide range of thermally induced movement in a building superstructure, expansion joint covers on subsurface structures nonetheless experience cyclical bending stress, which subjects them to flexural failure. More commonly, end seams at joints and plane transitions rupture. Joint covers must be adequately supported to prevent distortion from water and earth pressure.

Design expansion joint covers to maintain continuity at all changes in plane. For example, a tunnel intersecting a cellar may have an expansion joint on all four sides. You will need to draw an isometric of the joint between the horizontal and vertical joints.

Expansion joints in horizontal surfaces should be raised on cant strips at least 1½ inches high. Cement cants are preferable to treated wood because (1) they don't require mechanical fasteners, and (2) they assure compatibility between slab and membrane (see Figure E.2). If a joint does leak, an elevated cant will limit the quantity of water entering the building. It will also relieve some of the pressure on the expansion joint cover. Design the

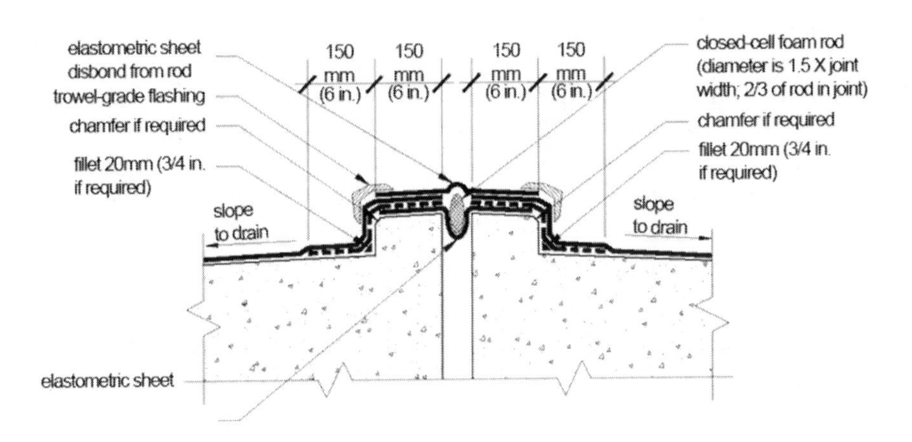

Figure E.2 (ASTM International)

elastomeric cover supports to accommodate flexure. The support should be inserted in the joint and designed to resist earth and water pressure. Supports should be round, solid, closed-cell sponge neoprene rods, and 25 to 33 percent larger than the joint width when installed. Tubes should conform to ASTM D1752 Type 1 (or D1056) with a 30-pound per cubic foot density. Solid rods are preferred to hollow tubes because tubes may serve as a conduit if the joint cover leaks. This would increase the difficulty of leak detection and location.

On horizontal surfaces of suspended slabs round (non-proprietary) supports should be installed on continuous neoprene or butyl hammocks to provide a vapor retarder and bitumen stop. The bitumen stop is mandatory when the membrane contains coal tar pitch or coal-tar-modified urethane. Fiberglass insulation may be installed in the joint to reduce heat loss, but it may keep the temperature of the vapor retarder below the dewpoint and cause condensation.

Proprietary rubber semicircular or quarter-circular units with tubes and bayonet extensions may be used in lieu of a rod to support the expansion joint cover. This detail eliminates a hammock and provides a more uniform surface to support the joint cover. Proponents claim that a hammock makes leak detection and identification more difficult. In my experience, a hammock is worth that risk because it provides redundancy. To be effective, it must be discharged to a floor drain or wet sink. Alternatively, the drain line may be equipped with leak detection instruments to alert maintenance personnel that the expansion joint cover has failed.

The use of pre-compressed impregnated foam joint fillers simplifies detailing joints. Where they are used with post-applied waterproofing systems, the use of angles to frame the joint walls will provide a means of constructing watertight terminations at the joint walls.

Authorities such as Charles Parise recommend installation of a sheet metal gutter under an expansion joint. Others, such as Stephen Ruggiero and Dean Rutila, believe this encourages occupants and maintenance workers to overlook problems. If a hammock/rod is omitted in favor of a proprietary joint cover support, you should consider using a sheet metal gutter under the joint to reduce the potential for catastrophic leaking (see Figure E.3).

When including gutters, design them with a one percent slope to drain and a V-shaped bottom to promote water flow and scouring. A piping system to an open drain is required. The top should be at least four inches below the slab to permit flushing and maintenance. The slab soffit should be provided with drips on both sides of the joint to prevent lateral migration of water.

Expansion joints in plaza wearing courses must be closer together than those in the structural slab. Structural expansion joints should generally be spaced between 100 and 200 feet, whereas joints in the wearing course should be spaced approximately 20 feet each way.

Calculating expansion joint width

As with many above-grade buildings, differential, rather than cumulative movement controls the location and size of expansion joints in below-grade structures. For buildings above-grade, exposed to the elements and the full range of ambient temperatures, expansion joint widths for thermally

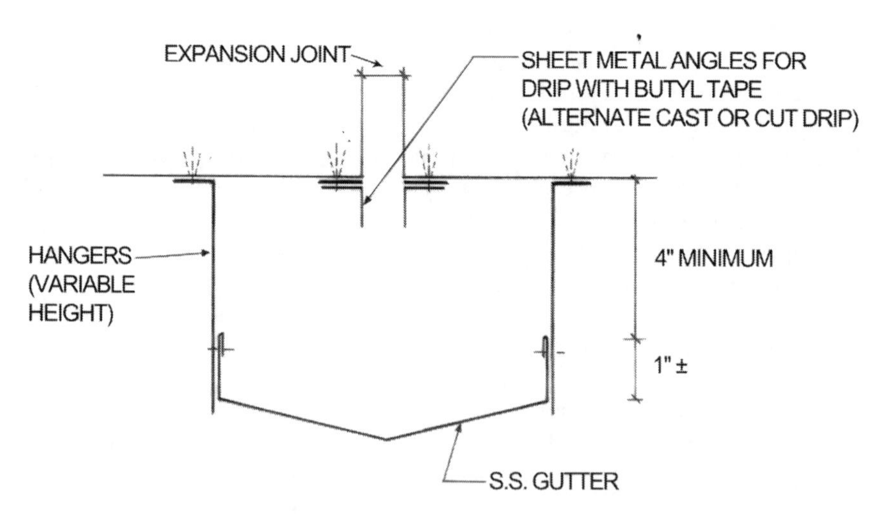

Figure E.3 Ceiling gutter detailed above should slope one percent to drain, with a V-shaped bottom to promote flow and scouring. The slab soffit should have drips on both sides of the joint to prevent lateral migration of water. Hangers should extend a minimum of four inches below the slab soffit to facilitate flushing and maintenance

induced cumulative movement can be calculated by using the formula: joint width $\varepsilon \times T \times L$, where ε is the coefficient of thermal expansion, T is the temperature differential and L is the length of the element. This formula is applicable to expansion joints in plazas, but inappropriate for joints in below-grade structures, where temperature differentials are not the dominant force that produces movement at "expansion" joints.

In below-grade structures, the amount of differential movement from settlement, flotation, and seismic forces is often greater than thermally induced cumulative movement. Temperature differentials are only significant during construction, before backfilling, and prior to insulating structural slabs under plazas.

The width of expansion joints in below-grade structures should be calculated by a geotechnical engineer (for settlement and flotation) and by a structural engineer (for seismic movement).

Generally, expansion joints in concrete structures should never be less than an inch wide when installed at the highest ambient temperature at which they will be exposed. Joints may be more than three inches wide in high seismic areas or where tidal-influenced groundwater levels fluctuate. Two-stage seals on the outer and inner faces are strongly recommended for all joints except those in plazas, where a single-stage joint seal is usually satisfactory. The outer part of the joint is formed by an elastomeric membrane carried over a flexible, bulb-shaped support. The inner part of the joint should be filled with a pre-compressed expanding open-cell foam saturated with modified acrylic.

Notes

1 According to Brent Anderson, P. E. Underground Waterproofing, 1983, Concrete at 50% RH will shrink 5/8-inch in 100 feet or 1/8-inch in 20 feet. Inch.
2 Ibid. Concrete cured for 7 days and exposed to 50% RH will experience 30% of its total shrinkage within 30 days and 85% in 6 to 9 months.

Appendix F

Waterproofing selection procedure

The seven flow charts in this appendix demonstrate formal design process. They are reprinted from U.S. Navy publication NAVFAC DM-1.4 Earth-Sheltered Buildings, Design Manual 1.4. Each algorithmic flow chart shows the steps to take in each of these sequential stages:

1. Evaluation of groundwater conditions.
2. Occupancy parameters.
3. Soil contaminants.
4. Structural requirements.
5. Construction procedures.
6. Product reliability.
7. Product performance.

These algorithms constitute a highly detailed elimination process, converging on the solution to the waterproofing problem. The first five charts (Figures F.1 through F.5) deal with technical problems. The last two charts (Figures F.6 and F.7) focus on product liability, contractual relationships, and product performance.

Figure F.6 is of particular interest. This algorithm puts each manufacturer through a rigorous investigation. The answers to one set of questions tell you whether to re-evaluate or test a product under consideration. The answers to another set of questions tell you if, for example, the manufacturer "endorses" the applicator's work, thus implicitly accepting responsibility for it. If not the task of assigning responsibility in event of a failure will be complicated, as each party seeks to escape blame and attribute the failure to others. Considering the professional malpractice risks in our super-litigious society, you should find Figures F.6 and F.7 very helpful for evaluating waterproofing products and their manufactures.

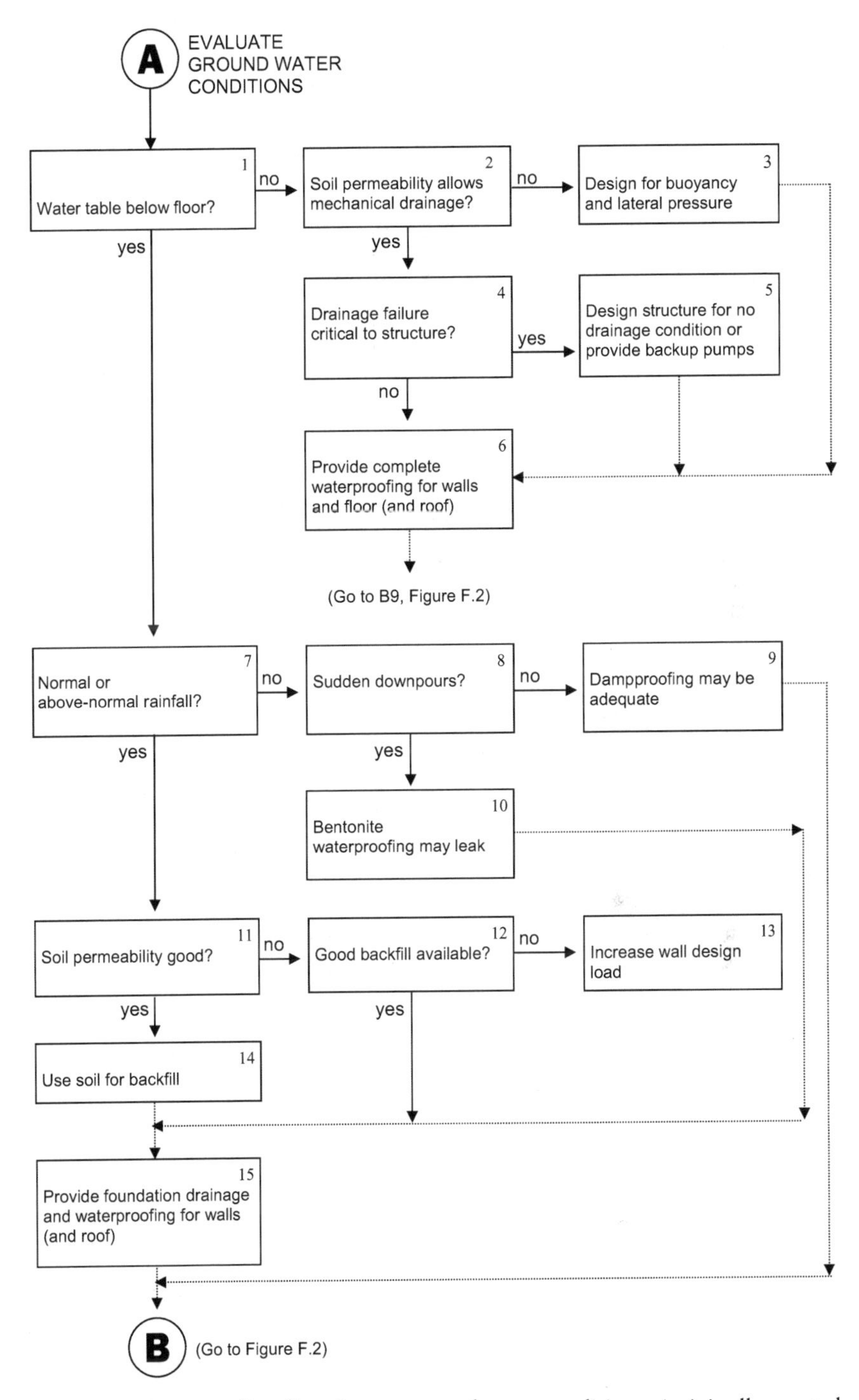

Figure F.1 Waterproofing flowchart – groundwater conditions. (originally created by Naval Facilities Engineering Command; Alexandria, VA)

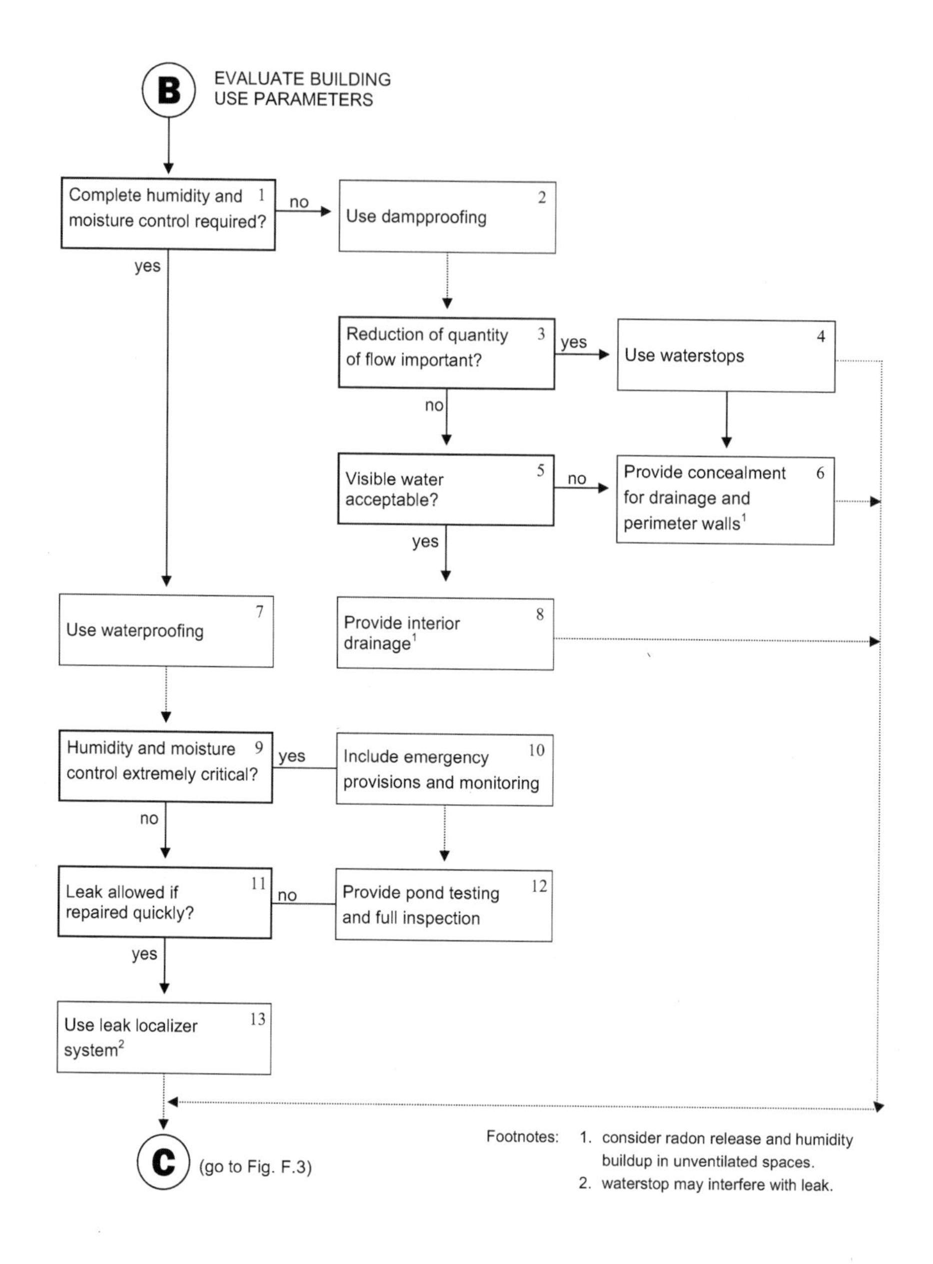

Figure F.2 Waterproofing Flowchart – Building Use Parameters. (originally created by Naval Facilities Engineering Command; Alexandria, VA)

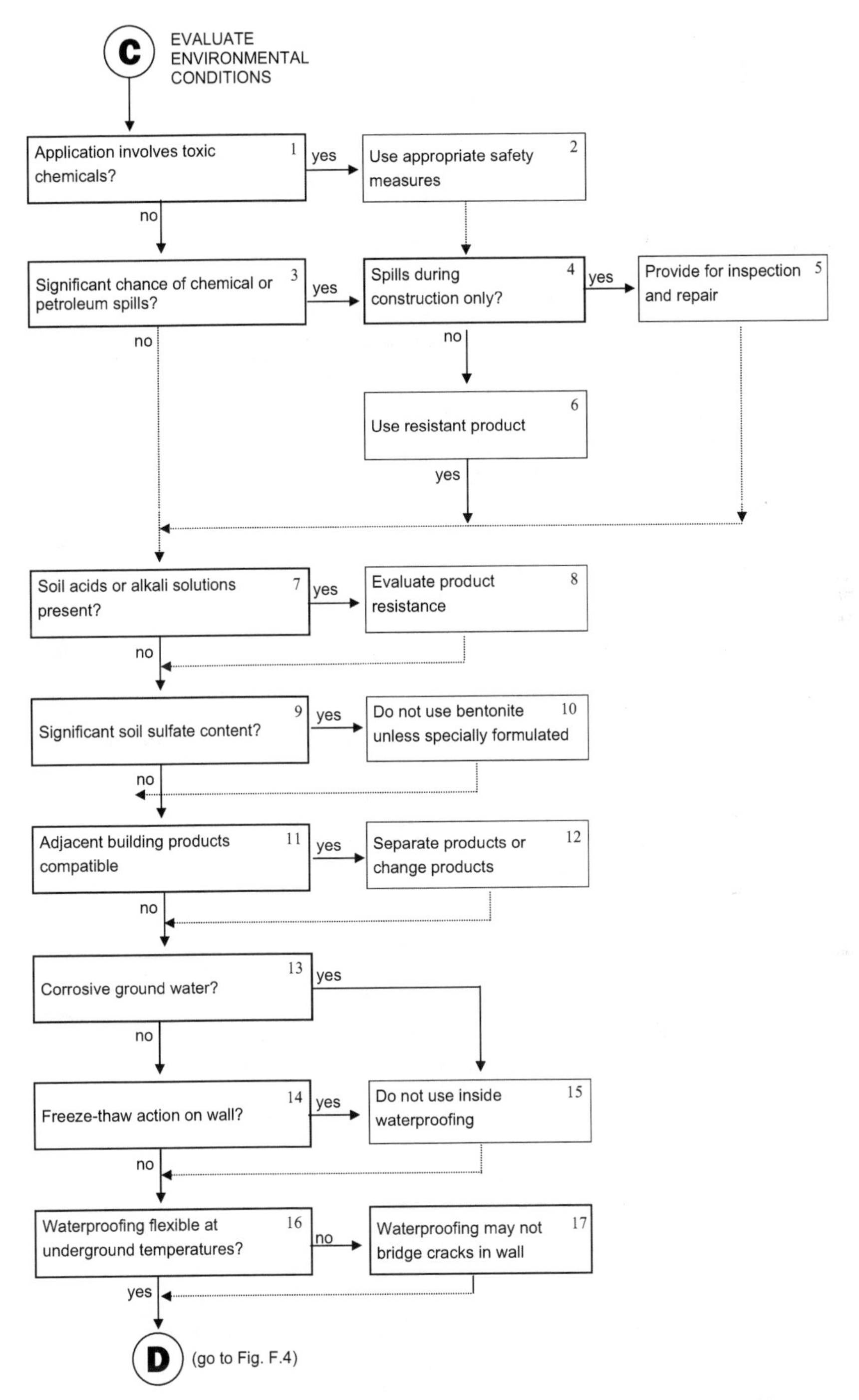

Figure F.3 Waterproofing flowchart – environmental conditions. (originally created by Naval Facilities Engineering Command; Alexandria, VA)

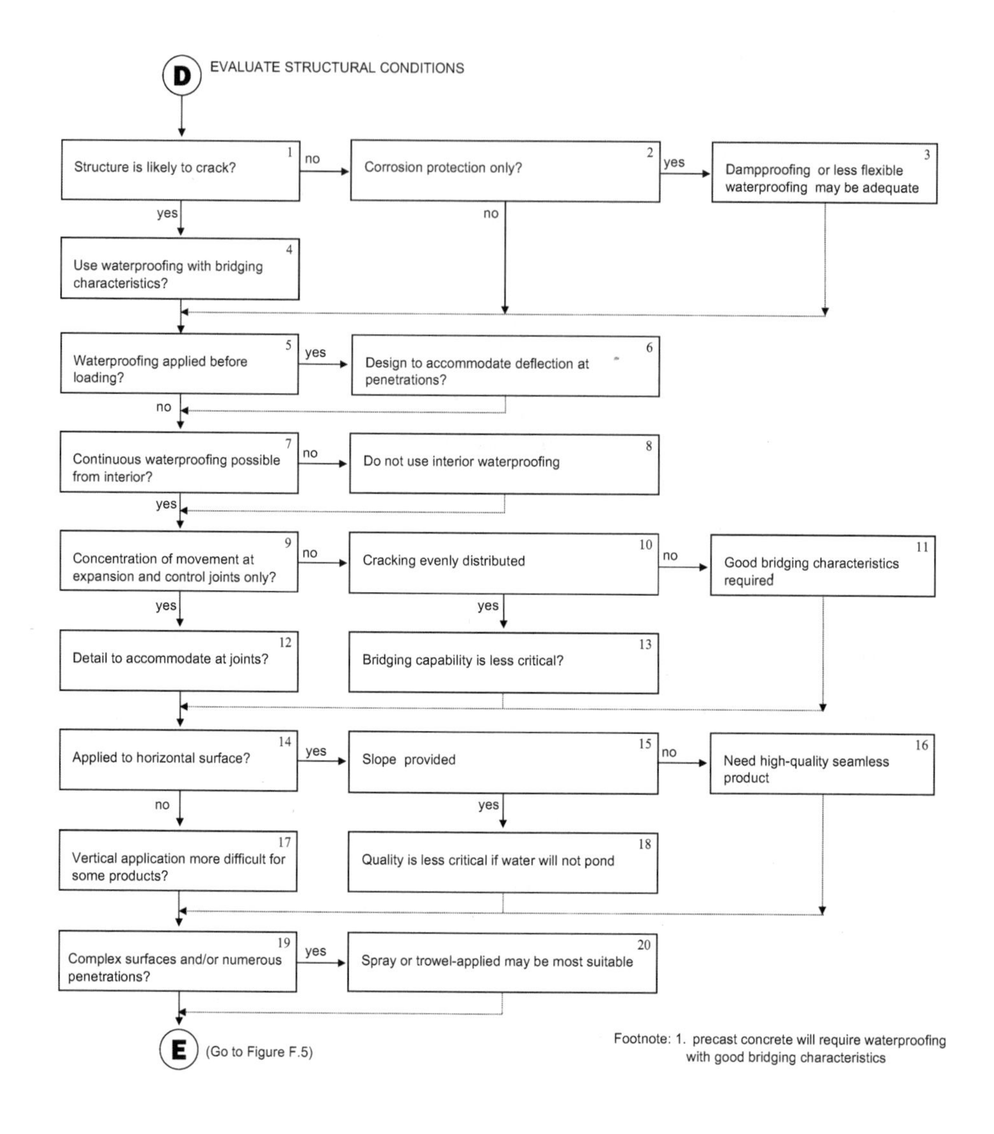

Figure F.4 Waterproofing flowchart – structural conditions. (originally created by Naval Facilities Engineering Command; Alexandria, VA)

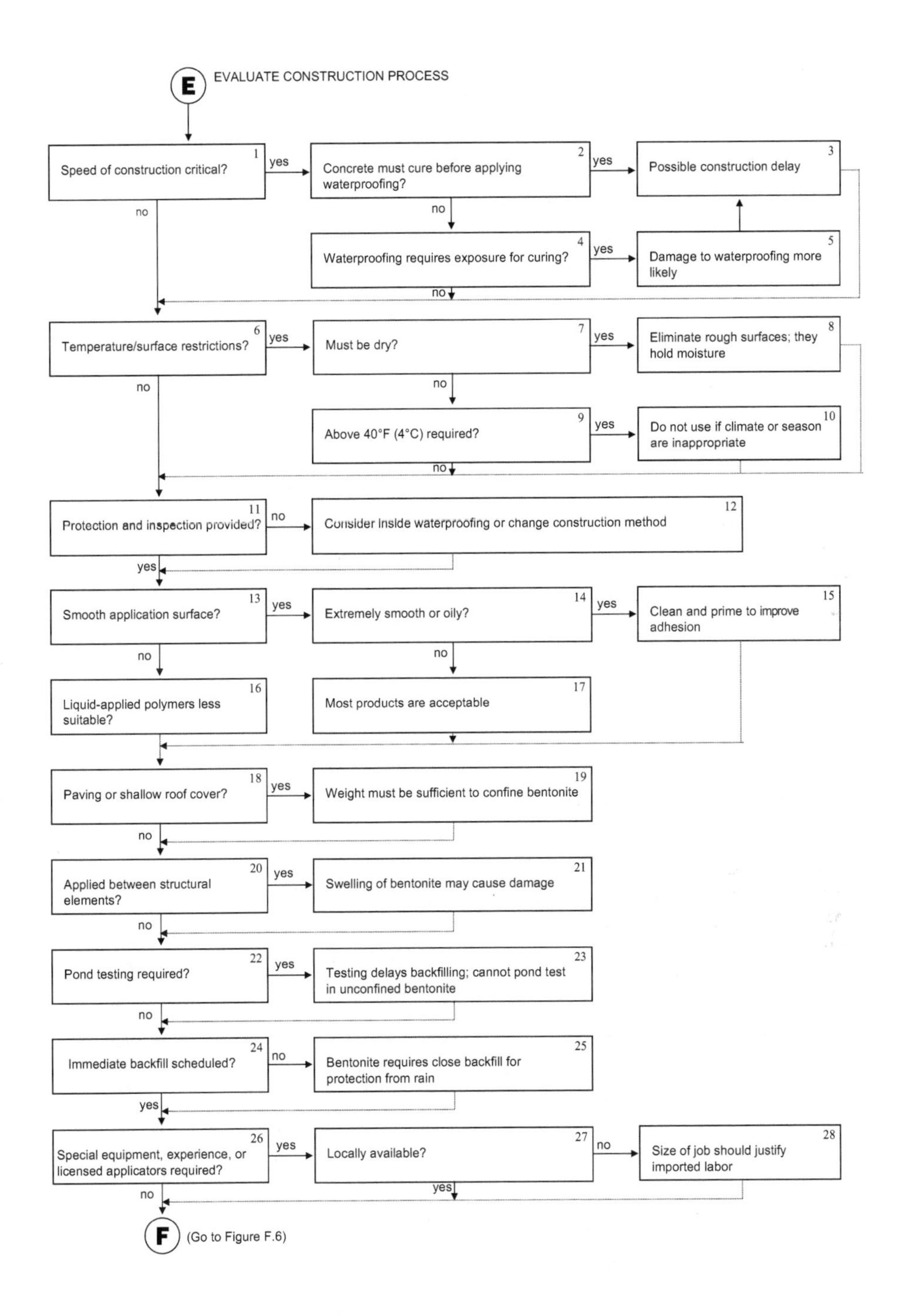

Figure F.5 Waterproofing flowchart – construction process. (originally created by Naval Facilities Engineering Command; Alexandria, VA)

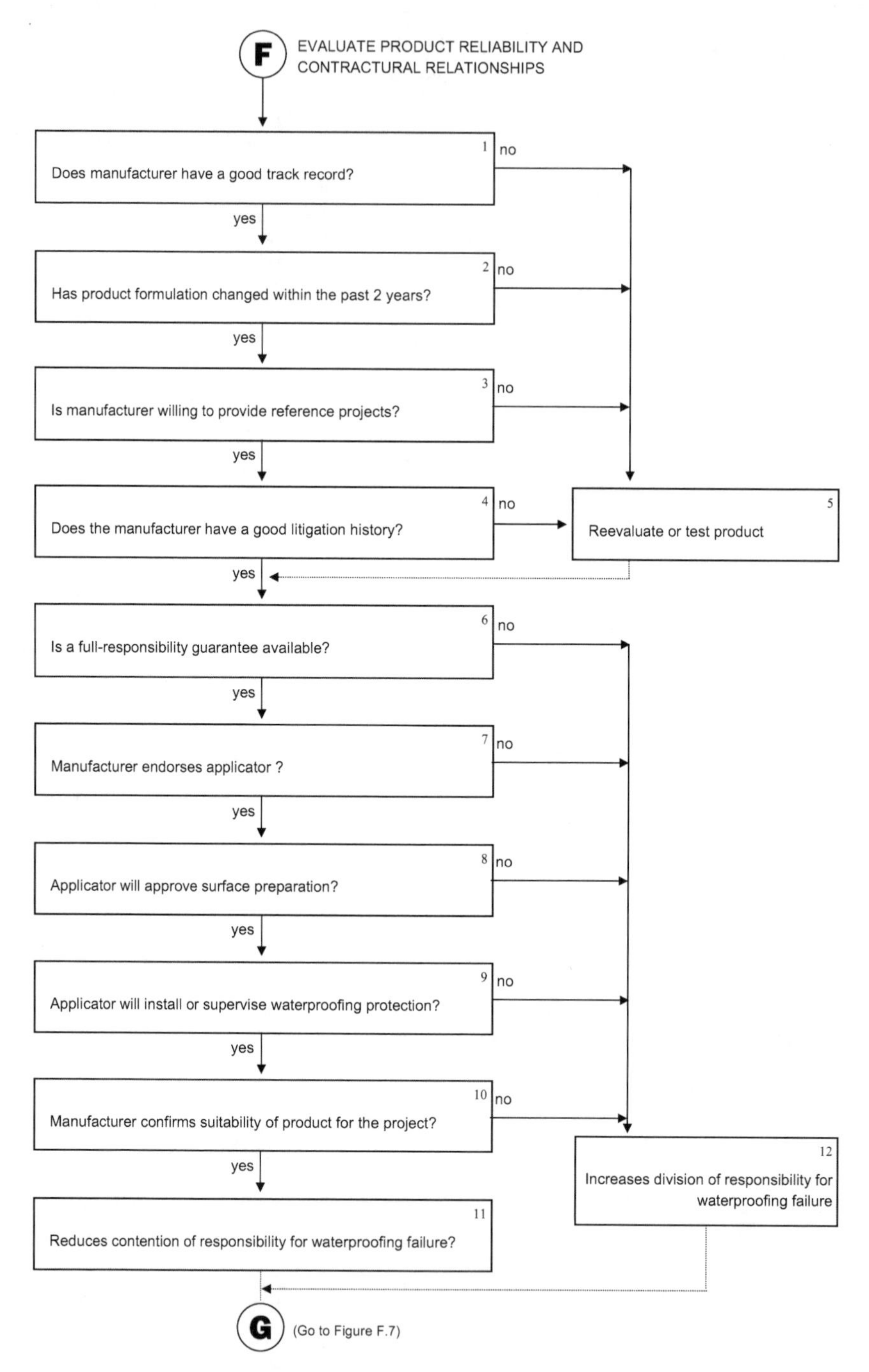

Figure F.6 Waterproofing flowchart – product reliability and contractual relationships (originally created by Naval Facilities Engineering Command; Alexandria, VA)

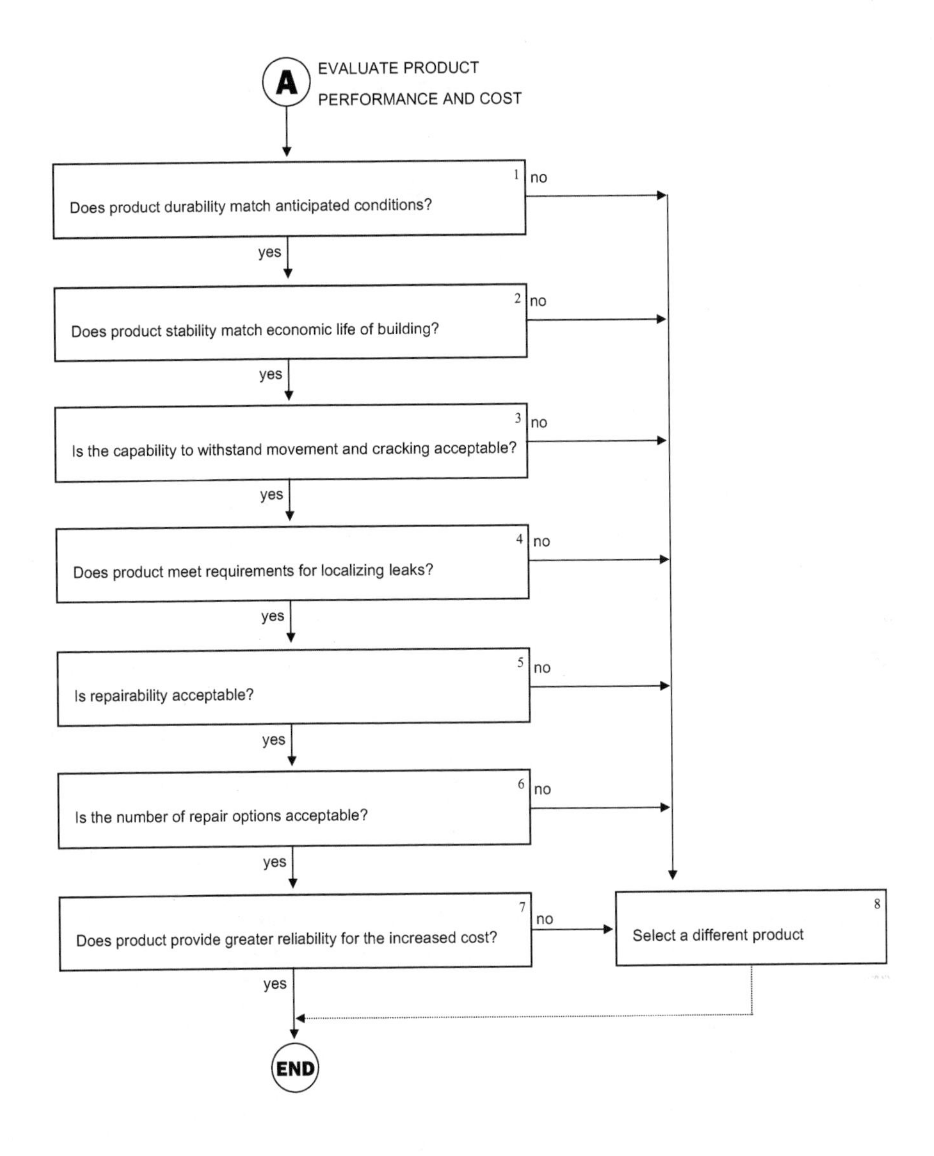

Figure F.7 Waterproofing flowchart – product performance and cost (originally created by Naval Facilities Engineering Command; Alexandria, VA)

Glossary of waterproofing terms

Admixture A material (other than water, aggregates, hydraulic cement, and fiber reinforcement) used as an ingredient of concrete or mortar, and added to the batch immediately before or during its mixing.

Asphalt A dark brown to black cementitious material whose predominating constituents are bitumens that occur in nature or are obtained in petroleum processing.

Asphalt felt An asphalt-saturated felt.

ASTM Formerly: American Society for Testing and Materials. Presently: ASTM International.

Backfill Soil or gravel placed adjacent to a foundation after excavation for an underground structure.

Backnailing Nailing along the top edge of vertically applied waterproofing sheets or felts.

Basement A habitable space enclosed by foundation walls that retain earth.

Bentonite (sodium montmorillonite) Granulated smectite clay that swells five to six times its original volume in the presence of moisture.

Bitumen (1) A class of amorphous, black or dark-colored, (solid, semi-solid, or viscous) cementitious substances, natural or manufactured, composed principally of high-molecular-weight hydrocarbons. Bitumen is soluble in carbon disulfide, and found in asphalts, tars, pitches, and asphaltenes. (2) A generic term used to denote any material composed principally of bitumen. (3) The two basic bitumens in the waterproofing industry are asphalt and coal tar pitch. Before application they are (a) heated to a liquid state, (b) dissolved in a solvent, or (c) emulsified.

Bituminous emulsion A suspension of minute globules of bituminous material in water or in an aqueous solution.

Blindside waterproofing Waterproofing applied to the surface of foundation formwork facing the excavation, resulting in its final location on the outside ("blindside") of the foundation wall. Also called *pre-applied.*

Boarding out The practice of applying a plywood facer to lagging or sheet piling to provide a smooth, plane substrate for waterproofing application.

Bond breaker A thin sheet installed in a joint or seam that divorces facing materials.

Bug holes Small regular or irregular cavities, usually not exceeding $^1/_2$- to $^5/_8$-inch in diameter, resulting from entrapment of air bubbles in the surface of formed concrete during placement and consolidation.

Buoyancy The vertical force causing the building to float when surrounded by water. $F_b = wAH$, where F_b is the force due to buoyancy, w is the specific weight of water (62.4 pounds per cubic foot), A is the area of the floor slab in square feet, H is the height of the water above the floor slab

Cant A triangular beveled strip placed or formed in an inside corner to modify the 90° angle to two 45° angles.

Capillarity The movement of a liquid in the interstices of soil or other porous material, due to surface tension, to an elevation above the general groundwater level.

Capillary break A membrane extended across the top of a wall footing to prevent capillary rise of water in the foundation wall.

Chamfer Either a beveled edge or corner formed in concrete work by means of a chamfer strip.

Chamfer strip Either a triangular or curved insert placed in an inside form corner to produce either a rounded or flat chamfer or to form a dummy joint. Also called *fillet*, and *skew back*.

Chemical grout Polyurethane-based hydrophilic liquids that react with water and change into an elastomeric solid.

Chimney drains Vertical strips of drainage materials, such as prefabricated drainage composites or porous fill, spaced along the foundation. Usually found on blindside waterproofing where outlets are provided at the bottom to underslab drainage systems.

Coal tar felt A felt saturated with refined coal tar.

Coal tar pitch A dark brown to black, solid cementitious material obtained as residue from the partial evaporation or distillation of coal tar.

Cold joint A plane of weakness in concrete caused by an interruption or delay in the casting operation. It causes the first batch to start setting before the next batch is added, resulting in a lack of bond between the batches.

Construction joint In concrete construction, a formed or assembled joint between two adjacent casting operations ("pours," "lifts," or "placements").

Contraction joint (Also called *control joint*). A formed, sawed, or tooled groove in a concrete structure to create a weakened plane and regulate the location of cracking resulting from the dimensional change of different parts of the structure.

Control joint See *contraction joint* and *cold joint*. Control joints are recessed, with "V" or "U"-shaped cross-sections detailed to shade cracks and thus minimize their aesthetic impairment.

Curing (1) The process through which concrete hardens after cement is mixed with water. (2) The process through which liquid-applied coatings harden into solid, elastomeric membranes.

Curing compound A liquid coating applied to the surface of newly cast concrete to retard the loss of mixing water. It promotes hydration of the cement. Pigmented curing compounds are also designed to reflect surface heat, which prematurely dries the concrete and impairs the cement hydration process.

Cut-back Solvent-thinned bitumen used in cold-process waterproofing and dampproofing adhesives, flashing cements, and coatings.

Dampproofing Treatment of a surface or structure to resist the passage of water vapor and water in the absence of hydrostatic pressure.

Deadman Concrete mass outside an excavation anchoring tie-backs against lateral earth pressure.

Deck The non-vertical structural surface to which the waterproofing system (including insulation) is applied.

Drainage composites Also called *prefabricated drainage composites* (PDC). Three-dimensional plastic sheets or fused plastic filaments formed to provide multidirectional flow with one or both sides covered with a geotextile.

Drainage course (percolation layer) A layer of washed gravel or manufactured drainage medium that allows water to filter through to a drain.

Drainage fill (1) Base course of granular material placed between floor slab and sub-grade to impede capillary rise of moisture. (2) Granular material placed on suspended slabs below grade or on plaza decks to promote drainage.

Dry (n.) A material that contains no more water than it contains at its equilibrium moisture content.

Emulsion A dispersion of fine particles or globules of a liquid in a liquid. Asphalt emulsions consist of asphalt globules, an emulsifying agent such as bentonite clay, and water.

EPDM (ethylene propylene diene terpolymer) A synthetic elastomer based on ethylene, propylene, and a small amount of a non-conjugated diene to provide sites for vulcanization. EPDM features excellent heat, ozone, and weathering resistance and low-temperature flexibility.

Equilibrium moisture content (1) Moisture content of a material stabilized at a given temperature and relative humidity. An expressed amount is the percentage of the material that is moisture. (2) The typical moisture content of a material in a particular geographical location.

EVT (equiviscous temperature) The temperature (± 25°F) (±1°C) at which the viscosity of a bitumen is appropriate for application. Viscosity units are generally expressed in centipoise or centistokes.

Expansion joint A formed or assembled joint designed to allow independent movement between adjacent building segments or components without stress transfer.

Fabric A woven cloth of organic or inorganic filaments, threads, or yarns.

Felt A flexible sheet manufactured by interlocking fibers through a combination of mechanical work, moisture, and heat; but without spinning, weaving, or knitting. Waterproofing felts are manufactured from vegetable fibers (organic felts), glass fibers (glass-fiber felts), or polyester fibers (synthetic-fiber mats).

Fin A narrow linear projection on a formed concrete surface, resulting from mortar flowing into gaps in the formwork. See Figure G.1.

Figure G.1 Fins

Finger cut A preformed felt or fabric membrane slit to fit around rebars. Also called *pig ear*.

Finished wearing surface See *wearing surface*.

Fishmouth In built-up and single-ply membranes, a defective, half-conical opening at a wrinkled edge, resulting from failure to adhere a sheet or felt to its substrate.

Flashing The system used to seal membrane edges at walls, expansion joints, drains, gravel stops, and other places where a membrane is interrupted or terminated. Base flashing covers the edges of a membrane. Cap flashing or counterflashing shields the upper edges of the base flashing.

Float finish A rough concrete surface texture obtained by finishing with a wood or carpet float. Also called *sidewalk finish*.

Fluid-applied elastomer An elastomeric material, fluid at ambient temperature, that dries or cures after application to form a continuous membrane. Such systems normally do not incorporate reinforcement.

Footing A continuous slab cast under a foundation wall, designed to transmit the wall's compressive load to the underlying foundation soil. Also used under isolated columns.

Footing drain A pipe drain, either porous or perforated, located on the inside or outside (or both) of a footing to draw down the surrounding groundwater to relieve hydrostatic pressure.

Form kick-out A bulge in the foundation where concrete formwork has been displaced laterally.

Form oil Oil applied to the interior surfaces of forms to facilitate release from the concrete when the forms are removed. See also *bond breaker* and *release agent*.

Foundation The concrete or masonry support designed as a wall or column footing, pier, caisson, or pile cap transmitting superstructure loads to soil or rock.

Geotextile polypropylene, polyester, or nylon fabrics. Can be woven or non-woven (spun-bonded). Manufactured to provide a narrow range of permeability.

Glass felt Glass fibers bonded into a sheet with resin. Can be impregnated in bituminous waterproofing, roofing membranes, and shingles during the manufacturing process.

Grade beam A below-grade concrete beam running between piles or pile caps or between exterior column footings, and functioning as a continuous support for the foundation.

Green roof A subsurface framed deck covered with earth and vegetation. Also called *vegetative roof*.

Groundwater level The elevation in soil where water is at atmospheric pressure. Also called *water table*.

Gunite A proprietary term for shotcrete.

HDPE (high density polyethylene) A chemically inert material with fair resistance to sunlight.

Honeycomb Voids left in concrete due to failure of mortar to effectively fill the spaces between coarse aggregate particles.

Honeycombing A defect in concrete construction, characterized by voids where the concrete should be a continuous solid. Honeycombing is usually caused by either concrete mix having been cast with improper slump or failure to vibrate the concrete mix properly during the casting operation. See also *rock pocket.*

Hydrostatic pressure Pressure exerted by stationary water, in all directions, against adjacent surfaces. Hydrostatic pressure is directly proportional to the depth of water; it equals the product of the water's density (62.5 pcf or 1001.25kg/m^3) times its depth (in feet or meters), yielding a figure in psf or kg/m^2.

Integral waterproofing A system to waterproof or densify concrete that uses additives, such as stearates or superplasticizers, rather than a membrane or coating.

Intermittent water pressure Variable hydrostatic pressure occurring where the groundwater table is within capillary reach of the foundation.

Invert The inside elevation of the bottom of a pipe. Also called *invert level.*

Lagging Wood planks spanning horizontally between soldier piles, transmitting lateral earth pressure to the soldier-pile flanges, against which the lagging planks are wedged. See Figure G.2.

Figure G.2 Lagging on soldier piles

Laitance A weak layer of cement and aggregate fines on a concrete surface.

Liquid-applied membrane (LAM) See *fluid-applied elastomer.*

Loose-laid membrane An unadhered, single-ply membrane anchored to the substrate only at edges and penetrations.

Membrane A flexible or semi-flexible waterproofing material whose primary function is to prevent passage of moisture.

Mud slab An unreinforced concrete slab at least two inches (50 mm) thick that serves as the substrate for membrane waterproofing for a slab-on-ground or wall footing.

Nailer A strip of wood or other fitting attached to or set in concrete, to facilitate nailed connections.

Negative-side waterproofing A waterproofing system in which the source of hydrostatic pressure and the water-resisting element are on opposite sides of the structural component.

One-on-one Installation of a multiple-ply membrane with each ply installed over the last (as opposed to shingled layers). It produces a membrane with a continuous film of inter-ply bitumen.

Parging A thin coat of mortar applied to rough masonry walls to create a smooth surface.

POE (point of entry) A drawing that shows all the penetrations in the foundation and pressure slab that usually require sleeves for utilities such as electricity, plumbing, and gas.

Pea gravel Screened gravel, with a majority of particles passing a 9.5 mm ($^3/_8$ inch) sieve, but retained on a 4.75 mm (No. 4) sieve.

Perched water table A water table elevated above the normal, free-water elevation by impervious, unsaturated soils separating local bodies of water at varying elevations.

Percolation layer A layer of washed gravel or manufactured drainage media that allows water to filter through to the drain. Also called *drainage course.*

Permeability (1) The capacity of a porous medium to conduct or transmit fluids. (2) The rate of flow of a liquid or gas moving through a barrier in a unit time, unit area, and unit pressure gradient not normalized for, but directly related to thickness. (3) The product of vapor permeance and thickness (for thin films, ASTM E96; for those over $^1/_8$ inch, ASTM E96 Usually reported in perm inches or grain/(h.ft^2 . in. Hg) per inch of thickness.

Permeance The rate of water vapor transmission per unit area at a steady state through a membrane or assembly, expressed in ng/(Pa.s.m^2) [grain/(ft^2 h .in.Hg)].

pH (1) The negative log of a hydrogen-ion concentration; a measure of acidity and alkalinity. (2) A measure of the relative acidity or alkalinity of water. A pH of 7.0 indicates a neutral condition. A greater pH indicates alkalinity, and a lower pH, acidity. A one-unit change in pH indicates a tenfold change in acidity and alkalinity.

Pinhole A tiny hole in a liquid-applied waterproofing membrane.

Plasticizer A material, often solvent-like, incorporated in a plastic sheet to increase its workability, flexibility, or extensibility. The most important use of plasticizers is in PVC membranes, where the type of plasticizer dictates the membrane's use.

Polyethylene A thermoplastic, high-molecular-weight organic compound, used in sheet form as a vapor retarder. See *HDPE.*

Ply A layer of felt or fabric in a waterproofing membrane. A four-ply membrane should have at least four plies of felt or fabric at any vertical cross section cut through the membrane.

Positive-side waterproofing A waterproofing system in which the water-resisting element and the source of hydrostatic pressure are on the same side of the structural component.

Pressure slab A cast-in-place, reinforced concrete floor slab structurally designed to resist uplift from hydrostatic pressure.

Prime coat (1) A coating applied to a surface to enhance the bond between a substrate and subsequent coats or sheets. (2) The first liquid coat applied in a multiple-coat system.

Primer A compatible coating designed to enhance adhesion.

Primer (bituminous) A thin liquid bitumen applied to a surface to improve adhesion of heavier applications of bitumen and to absorb dust.

Protection board See *protection course*

Protection course Semi-rigid sheet material placed on a waterproofing membrane to protect it against damage during subsequent construction, and to provide a protective barrier against compressive and shearing forces induced by materials placed on or above it.

Puncture resistance An index of a material's or system's ability to withstand pressure from a sharp object without penetration.

PVC (polyvinyl chloride) A synthetic thermoplastic polymer prepared from vinyl chloride. PVC can be compounded into flexible and rigid forms through the use of plasticizers, stabilizers, filler, and other modifiers. Rigid forms are used in pipes. Flexible forms are used in sheets for single-ply roofing and waterproofing membranes.

Raker A diagonal brace supporting a soldier pile against lateral earth pressure. Rakers have been generally replaced by tie-backs, which do not clutter an excavation. See Figure G.3.

Reglet (1) A continuous groove or slot in a wall, designed to receive other components – e.g., flashing gaskets or anchors. (2) A continuous prefabricated metal or plastic device containing a groove, slot or recess that can be cast into (as a form) or mounted onto a building component surface.

Reinforcing Generally, one or more strips of membrane, felts or fabrics, installed at corners and over construction joints.

Release agent Material used to prevent bonding of concrete to a surface. See also *bond breaker* and *form oil.*

Figure G.3 Rakers

Rock anchor Reinforcing bars grouted into sockets that are drilled into foundation rock and extended into foundation slabs – e.g. isolated column footings to resist upward hydrostatic pressure.

Rock pocket A defect in concrete construction: a severe form of honeycombing in which large pieces of aggregate are exposed in voids formed in cast-in-place concrete foundation walls. This is caused by segregation and failure to vibrate the concrete enough to ensure that homogeneous distribution of cement, fines, and aggregate forms a continuous cross section through the wall. See also *honeycombing*.

Rupture A membrane tear resulting from tensile stress.

Scrim Woven open fabric, usually made from glass or polyester fibers, used to reinforce PVC waterproofing sheets.

Sheathing The material forming the contact face of forms. Also called lagging or sheeting.

Sheet pile concrete or wood planks, often splined, or roll-formed interlocking steel sections driven vertically to support the sides of an excavation. Also called *sheet piling*.

Shotcrete Mortar or concrete pneumatically projected at high velocity onto a surface.

Slab-on-ground An unframed (i.e. non-structural), cast-in-place concrete slab, supported by continuous bearing stress on the foundation soil, whose top surface is partly or wholly below the adjoining grade. Also called *slab-on-grade.*

Slurry A viscous, liquid mixture of water and suspended clay or cement particles.

Slurry trench A trench whose sides are retained by the lateral hydrostatic pressure of a slurry pumped in as the excavation proceeds.

Slurry wall A foundation wall constructed in a slurry trench. When excavation of the slurry trench is completed, concrete is deposited via the tremie process, with the concrete mix displacing the slurry from the bottom and ultimately filling the entire slurry trench with the permanent foundation wall.

SOE (support of excavation) the earth retaining system at an excavation. Also called shoring wall.

Soldier beams Steel H or round section piles, usually driven at equal spacing along a building line and designed either as vertical cantilevers or as horizontally supported structural members resisting lateral earth pressure around an open excavation. Supports may be rods and anchors used as tiebacks or diagonal rolled steel sections. Called "soldier beams" because of their resemblance to a line of soldiers at attention.

Soldier pile A more technically accurate, but lesser used field term for soldier beam.

Split A membrane rupture resulting from tensile-stress failure.

Structural slab (1) A cast-in-place concrete deck over an occupied space spanning between beams, bearing walls, or other supports. (2) A reinforced, cast-in-place concrete slab with foundation soil used as a bottom form, but designed to resist bending caused by upward hydrostatic pressure.

Substrate The surface upon which films, treatments, adhesives, sealants, membranes, and coatings are applied.

Suspended slab See *structural slab*, definition (1).

Tanking (British) Waterproofing both the foundation and slab-on-ground and joining them.

Telltale A stake or other surface marker indicating an underground drain or cleanout.

Thixotropy The property of a gel to become fluid when disturbed or applied; often used to describe the quality of "non-sag" materials.

Tiebacks (1) Rods or cables anchored into rock or concrete deadmen outside an open excavation that provide temporary restraint for walers, eliminating rakers or cross-bracing struts, thus freeing the excavated space from obstacles to construction equipment. (2) A rod fastened to a deadman, a rigid foundation, or either a rock or soil anchor to prevent lateral movement of formwork, sheet pile walls, retaining walls, bulkheads, etc. Also called *soil nail.* See Figure G.4.

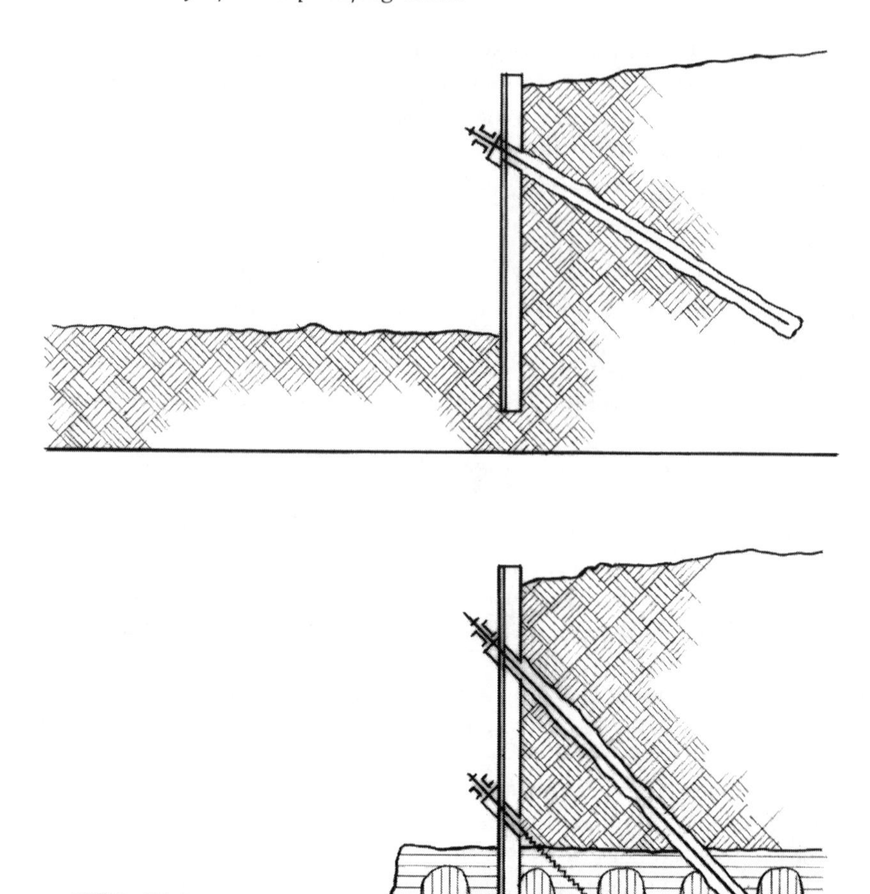

Figure G.4 Tiebacks in earth and into rock

Tie-off The transitional seal used to terminate a waterproofing application (1) at the top or bottom of flashing, or (2) by forming a watertight seal with the substrate, membrane, or waterproofing system.

Traffic surface A surface exposed to traffic, either pedestrian or vehicular. Also called *finished wearing surface*.

Vapor pressure Pressure exerted by a vapor in equilibrium with its solid or liquid form.

Vapor retarder A material that retards the flow of water vapor. Sometimes called a *vapor barrier*.

VOC (volatile organic content) Solvents that participate in the formation of reactive pollutants, particularly ozone, when released into the atmosphere. VOCs contribute to smog, cause respiratory problems, and damage crop yields. Maximum levels are established by the EPA and local regulatory bodies. Federal regulations currently allow the following carbon compounds to be emitted to the atmosphere during the application of and/or subsequent drying or curing of coatings: carbon monoxide, carbon dioxide, carbonic acid, metallic carbides or carbonates, ammonium carbonate, methane, 1,1,1,-trichlor ethane, methyl chloride and trichlorotrifluoroethane.

Wale A horizontal structural member that resists lateral soil pressure around an excavation during construction before the permanent foundation wall is constructed. Also called *waler* or *whaler*.

Water table The elevation (level) of soil where water is at atmospheric pressure. Also called *groundwater level*.

Waterproofing Treatment of a surface or structure to prevent passage of water under hydrostatic pressure.

Waterstop A preformed material placed between concrete pours to prevent passage of water through a joint.

Water-vapor transmission (WVT) The rate of water-vapor flow (under the conditions of a specified test such as ASTM E96) through a unit area of a material, in a unit of time between the material's two parallel surfaces. Customary units are grains/(h. ft^2). Also called *water-vapor transmission rate*.

Wearing surface A surface exposed to traffic, either pedestrian or vehicular. Also called *finished wearing surface*.

Well points Small-diameter wells, driven or jetted into soil surrounding an excavation and connected to a header pipe as part of a well-point system designed to drain water from an excavation. Well points lower the water table and relieve a building's foundation walls of hydrostatic pressure.

Bibliography

Also see Appendix D: Abstracts of ASTM standards relating to waterproofing

ACI 224.1R-93 *Causes, Evaluation and Repair of Cracks in Concrete Structures*, American Concrete Institute, Detroit, MI 1998

ACI Guide 515.1R79 *A Guide to the Use of Waterproofing, Dampproofing, Protective and Decorative Barrier Systems for Concrete.* American Concrete Institute. Detroit. MI (Revised 1984)

ALS Home Improvement Center *Waterproofing Basements*

Anderson, Brent. *Underground Waterproofing.* 1982. Earth Integrated Technics, Inc 1983

Anderson, Brent. "Waterproofing & the Design Professional." *The Construction Specifier* March 1986

Aquafin, Inc. Waterproofing, Vapor Emission Control, Restoration and Protection Systems 2004

ASTM C836 "High Solids Content Cold Liquid-Applied Elastomeric Waterproofing Membrane for Use with Separate Wearing Course"

ASTM C981 "Standard Guide for Design of Built-up Bituminous Membrane Waterproofing Systems for Building Decks." Section 13. "Insulation"

ASTM D1557-12e1 "Standard Test Methods for Laboratory Compaction Characteristics of Soil Using Modified Effort"

ASTM D1557-12e1, "Standard Test Methods for Laboratory Compaction Characteristics of Soil Using Modified Effort"

ASTM D4263 "Standard Test Method for Indicating Moisture in Concrete by the Plastic Sheet Method"

ASTM D449 "Standard Specification for Asphalt Used in Dampproofing and Waterproofing"

ASTM D5295 "Standard Guide for Preparation of Concrete Surfaces for Adhered (Bonded) Membrane Waterproofing Systems"

ASTM D5957 "Standard Guide for Flood Testing Horizontal Waterproofing Installations"

Babcock, Warner K. "Cementitious Membranes: Understanding the True Scope." *The Construction Specifier*. April 1982

BBZ USA, Inc. *Injection Materials and Waterproofing Technology* 2002

BRE *Basement Construction and Waterproofing* GBG 72 Parts 1 and 2 2007

British Standard RS8012, "Code of Practice for Protection of Structures against Water from the Ground." 1990

Byrd, Stacy, "Stopping Leaks with Engineered Bentonite Grout," *Specifier* March 2006

Cedergren, Harry R. *Seepage, Drainage, and Flow Nets* 3rd Edition. John Wiley & Sons. New York. 1997

Cedergren, Harry R. "Seepage Requirements of Filters and Pervious Bases." *Transactions of the American Society of Civil Engineers.* January 1962: 1099–1113

Committee on Earth Retaining Structures of the Geo-Institute of the American Society of Civil Engineers. *Guidelines of Engineering Practice for Braced and Tied-Back Excavations* (Geotechnical Special Publication (GSP) No. 74). Reston, VA 1997

Cornell, Russell W. "Waterproofing is as Waterproofing Does." *The Construction Specifier* December 1970

Construction Specifications Institute *Spectext* de Neef Inc. Construction Chemicals 2003

Dworkin, Joseph F. "Waterproofing Below Grade – Materials. Methods and Controls." *The Construction Specifier* March 1990: 44–52

Engel. Joseph. *Cold Applied Fluid Elastomeric Waterproofing Membranes.* Building Deck Waterproofing. ASTM STP 606. J. Panek, ed. ASTM. Philadelphia. 1976: 204–217

Estenssoro, Luis and William Perenchio. "Failures of Exterior Plazas." *The Construction Specifier* January 19911/91: 87–92

Friedberg. M. Paul. "Roofscape." *Architectural Engineering News* December 1969

Garden, G. K. "Roof Terraces." *Canadian Building Digest* CBD 75. National Research Council March 1966

Gaul. Robert W. "Preparing Concrete Surfaces for Coatings." *Concrete International* July 1984

Gibbons, Daniel and Jason Towle. "Waterproofing Below-Grade Shotcrete Walls." *The Construction Specifier*. March 2009

Griffin. C.W. and R.L. Fricklas. *The Manual of Low-Slope Roof Systems.* McGraw Hill. 3rd ed. 1996

Grimm, Clavford T. "Delamination of Masonry Pavements." *TMS Journal* February 1994

Haisley, (October 2001) Phil, *Pushing the Building Envelope...Below Grade, RCI Interface* Planert, Thomas, Substarte Testing), correspondence, 2015

International Concrete Repair Institute *Guide for Selecting and Specifying Materials for Repair of Concrete Surfaces.* Sterling. VA 1996

Kanarowski, Stanley M. *Evaluation of Bentonite Clay for Waterproofing Foundation Walls Below Grade.* CERL Technical Report M-93. May, 1975

Kessi, Alfred. "Negative Side Waterproofing." *Concrete Repair Bulletin International* Concrete Repair Institute. Sterling. VA September/October 1996

Kidder, Frank E and Harry Parker. *Kidder-Parker Architects' and Builders' Handbook.* Waterproofing for Foundations: 2123–2131. John Wiley & Sons. Eighteenth Edition 1945

Kubal, Michael K. *Waterproofing the Building Envelope.* McGraw-Hill Inc. New York. NY. 1993

Labs, Kenneth, John Carmody, University of Minnesota. Underground Space Center. *Building Foundation Design Handbook.* U.S. Dept. of Energy, Martin Marietta Energy Systems, Inc 1988

Lawrence. Dorothy. Lectures in California: October 24. 1990 and April 24. 1995

Lawrence, Dorothy and Charles O. Pratt. "Chapter 4 Waterproofing Barrier

Systems." Revised February 21, 1994 from *ACI Guide 515. 1R79. A Guide to the Use of Waterproofing, Dampproofing, Protective and Decorative Barrier Systems for Concrete.* American Concrete Institute Detroit, MI

Laaly. H.O. *The Science and Technology of Traditional and Modern Roofing Systems.* Griffin Printing. Glendale. CA 1993: Chapter 23

Macnab, Alan. *Earth Retention Systems Handbook*, McGraw-Hill, 2002

Maslow, Philip. *Chemical Materials for Construction.* Structures Publishing Co. Farmington, MI. 1974

Merriman, Thaddeus and T.H. Wiggin, eds. *American Civil Engineering Handbook.* John Wiley 5th Edition. February 1930: 1643

Monroe, D.C. *Reflective Cracking and Cold. Liquid-Applied Elastomeric Deck Coating and Membrane Systems: Practical Considerations from Field Observations* Building Deck Waterproofing. ASTM STP1084. L.E. Gish, ed. ASTM. Philadelphia. 1990: 121–130

National Roofing Contractors Association *The NRCA Roofing & Waterproofing Manual Third Edition.* 1989. Rosemont. IL

NAVFAC DM-1.4 *Earth Sheltered Buildings.* Dept. of the Navy. Alexandria, VA Government Printing Office. Washington, DC. March 1984

Oak Ridge National Laboratory. United States Department of Energy. "Waterproofing Considerations and Materials." ORNL/SUB-7849/04. September 1981

Parise, C.J. *Architectural Considerations in Plaza Membrane Waterproofing Systems.* Materials and Research Standards. ASTM Vol 11. No. 5. May 1971: 19–22

Pashina, Keith. "Coating Concrete." *The Construction Specifier.* December. 1993: 86

Polygem, Inc. Waterproofing Crack Injection Techniques

Pratt, Charles O. *Waterproofing – Problems in Terminology.* Building Deck Waterproofing. ASTM STP 1084. L.E. Gish, ed. ASTM. Philadelphia. 1990: 137–142

Ruggiero, Stephen S. and Dean A. Rutila. "Plaza Waterproofing Design Fundamentals." *The Construction Specifier.* January 1991

Ruggiero, Stephen S. "Buried Buildings." *Progressive Architecture.* June 1993

Ruggiero, Stephen S. and Dean A. Rutila. *Principles of Design and Installation of Building Deck Waterproofing.* Building Deck Waterproofing. ASTM STP 1084. L.E. Gish, ed. ASTM. Philadelphia. 1990: 5–28

Simmons, H. Leslie. *Repairing and Extending Weather Barriers.* Van Nostrand Reinhold. New York. 1989

Stilling, Robert G. "Waterstops – Materials and Configurations with Illustrated Test Results." Meeting Association of Conservation Engineers. North Platte, NE. November 8, 1967

"Technics: Roofs for Use." *Progressive Architecture.* July 1990

Tobiasson, Wayne. "Chapter 16 General Considerations for Roofs" from *Moisture Control in Buildings.* ASTM Manual 18. Heinz R. Trechsel, ed. Philadelphia. 1994

U.S. Army Corps of Engineers *Method of Test for Water Permeability of Concrete.* CRD-C 48-73. Revised December 1973

Webac Corp. *The Right Grout – Hydrophobic/Hydrophilic?* 2004

Index

9781032922799